On the cover. A typical neighborhood (muhallah) in Baghdad, 15 December 2006. U.S. Army.

JOINT SPECIAL OPERATIONS
JSOU
UNIVERSITY

Block by Block: Civic Action in the Battle of Baghdad

January–November 2006

Adrian T. Bogart III

JSOU Report 07-8
The JSOU Press
Hurlburt Field, Florida
2007

Comments about this publication are invited and should be forwarded to Director, Strategic Studies Department, Joint Special Operations University, 357 Tully Street, Alison Building, Hurlburt Field, Florida 32544. Copies of this publication may be obtained by calling JSOU at 850-884-2765; FAX 850-884-4732.

****

The Strategic Studies Department, JSOU is currently accepting written works relevant to special operations for potential publication. For more information please contact Mr. Jim Anderson, JSOU Director of Research, at 850-884-1569, DSN 579-1569, james.d.anderson@hurlburt.af.mil. Thank you for your interest in the JSOU Press.

****

ISBN 1-933749-05-9

Recent Publications of the JSOU Press

Theoretical Perspectives of Terrorist Enemies as Networks, October 2005, Robert G. Spulak, Jr. and Jessica Glicken Turnley

Logistic Support and Insurgency, October 2005, Graham H. Turbiville, Jr.

Dividing Our Enemies, November 2005, Thomas H. Henriksen

The War on Terrorism, December 2005, James A. Bates

Coast Guard SOF, February 2006, Gary R. Bowen

Implications for Network-Centric Warfare, March 2006, Jessica Glicken Turnley

Narcoterrorism in Latin America, April 2006, Alvaro de Souza Pinheiro

The Changing Nature of Warfare, the Factors Mediating Future Conflict, and Implications for SOF, April 2006, John B. Alexander

Civil-Military Operations and Professional Military Education, May 2006, James F. Powers, Jr.

Blogs and Military Information Strategy, June 2006, James Kinniburgh and Dorothy Denning

2006 JSOU/NDIA SO/LIC Chapter Essays, June 2006

One Valley at a Time, August 2006, Adrian T. Bogart III

Special Operations Aviation in NATO, September 2006, Richard D. Newton

Beyond Draining the Swamp: Urban Development and Counterterrorism in Morocco, October 2006, Stephen R. Dalzell

Filling Special Operations Gaps with Civilian Expertise, December 2006, James F. Powers, Jr.

Educating for Strategic Thinking in the SOF Community, January 2007, Harry R. Yarger

The Israeli Approach to Irregular Warfare and Implications for the U.S., February 2007, Thomas H. Henriksen

Psychological Operations: Learning Is Not a Defense Science Project, March 2007, Curtis D. Boyd

2007 JSOU and NDIA SO/LIC Division Essays, April 2007

Hunting Leadership Targets in Counterinsurgency and Counterterrorist Operations, June 2007, Graham H. Turbiville, Jr.

Executive Report, JSOU Second Annual Symposium (30 April–3 May 2007)

A Theory of Special Operations, October 2007, Robert G. Spulak, Jr.

To the 238 great soldiers who gave their lives during the United States (U.S.) 4th Infantry Division's fight in the Battle of Baghdad. They came from 9 of the 10 divisions across the U.S. Army and fought proudly as one. They fought with courage and honor, and they died as warriors, strong and true. They serve as an enduring testament to the strength of our country and an ever present reminder of the cost of freedom. Each and every one is an American hero.

Contents

Foreword

This monograph describes one facet of the Battle for Baghdad during the period January through November 2006. The story is based on the recollections, notes, and reports of the author, who served with the Multi-National Division, Baghdad (MND-B) as the G9—the principal staff officer responsible for civic action, Special Operations Forces integration, and counterinsurgency training. In this timeframe MND-B treated civic action as a maneuver function inherent to its operations, and it employed task-organized combat forces to conduct Phase IV (Stability Operations) and Phase V (Enable the Civil Authority) in order to achieve U.S. and Iraqi military objectives.

The sources for this report have since been declassified by the U.S. 4th Infantry Division. This division assumed command of the MND-B from the U.S. 3rd Infantry Division on 7 January 2006. The 3rd Infantry Division had just completed a historic year in Baghdad ushering in the elections for the national government to establish democracy in Iraq.

For the MND-B, the mission at hand was to secure Baghdad and provide the opportunity for the newly elected government to establish self-rule over the sovereign state of Iraq. To do so, the MND-B counterinsurgency operation assumed two components, security operations and civic action. The civic-action program was centrally planned with decentralized execution to accommodate the variances in the operational environment throughout the MND-B area of responsibility. This report gives a sense of the extensive efforts made by the MND-B troops to assist the local population in Baghdad while supporting command objectives. The kinds of accomplishments and the methods employed provide valuable insight for others who must conduct operations in similar circumstances.

On 15 November, the 4th Infantry Division transitioned command and control of MND-B to the U.S. 1st Cavalry Division. Joint Special Operations University is pleased to provide LTC Adrian T. Bogart's experiences in the Battle for Baghdad.

Michael C. McMahon, Lt Col, USAF
Director, JSOU Strategic Studies Department

About the Author

Lieutenant Colonel Adrian T. Bogart III is a Special Forces officer who has served in U.S. Army, joint, and combined organizations in command and staff assignments with special operations and general purpose forces. His military career started as a cadet at the Virginia Military Institute (VMI), commissioned upon graduation as an infantry officer with his first tour of duty in a mechanized infantry battalion in Germany. He left the infantry for Special Forces with assignments as a detachment commander, company executive officer, and company commander. He also served as a Special Forces staff officer on the Army Staff and was one of the founders of the Department of Defense domestic consequence-management capability, building the civil support teams and establishing the U.S. Northern Command.

LTC Bogart at the Crossed Sabers Memorial in Baghdad's Green Zone, Iraq's political and diplomatic center.

After his assignments in the Pentagon, LTC Bogart was the deputy C9 chief of operations for the Coalition Force Land Component Command (CFLCC) for Operation Iraqi Freedom serving in Kuwait and Iraq (2002 and 2003), later moving to 3d Special Forces Group where he served as the J5 and chief of staff/executive officer for the Combined Joint Special Operations Task Force Afghanistan (CJSOTF-A), Operation Enduring Freedom (2003 and 2004). LTC Bogart served as the G9 of the 4th Infantry Division deployed for Operation Iraqi Freedom 05-07 (fiscal year 2005–2007). Arriving 5 December 2005 and departing 15 November 2006, he was responsible for Special Operations Forces integration, civic action, and counterinsurgency training. His degree from VMI is in civil engineering, and he is a graduate of the U.S. Army Command and General Staff College, Army Force Management School, Defense Strategy Course, Joint Forces Staff College, and other military schools.

Acknowledgments

Those Who Bring Victory, These Quiet Few

I do not fear an army of lions, if they are led by a lamb.
I do fear an army of sheep, if they are led by a lion.

— Alexander the Great

This program was executed by some of the most gifted and talented combat leaders in the U.S. Army. Through the days and nights, across a very lethal battlefield, under extreme stress and in constant contact with the enemy, America's soldiers provided good governance, essential services, and economic development to a 16,664 square-mile area of operations, encompassing four Iraqi provinces, and including the capital city of Baghdad. These American soldiers achieved the impossible by building trust and confidence with a people who had long lost all hope for their future.

The success achieved from January to November 2006 during the Battle of Baghdad was forged by many officers and noncommissioned officers assigned to the Multi-National Division, Baghdad (MND-B). These individuals' personal courage and strength of will moved the concept of operations into reality, which came in the form of renovated schoolhouses, functioning agriculture cooperatives, public works improvements, medical care for thousands, and hundreds of other civic projects. Special acknowledgements follow:

a. Brigadier General David Halverson, U.S. Army, was the driving force that brought interagency and host national efforts in line with the MND-B's civic-action priorities. His direct guidance to the brigades was the synchronizing catalyst across the force. Without his strength, we would have failed.

b. Lieutenant Colonel Michael J. Repetski, U.S. Army, was the prime mover in understanding the operational environment. He brought insight into the inner workings of the Iraqi society and was able to make the division see what they did not want to see. He is the epitome of the warrior-diplomat.

c. Lieutenant Christine Kukla, U.S. Army Reserve, was the strength of the G9 43-member staff; she kept it going with a steady and sure hand. She led the staff towards the success achieved. She stood above the rest, a great warrior who ensured it all got done.

d. Lieutenant Colonel Robert A. D. Sauers, U.S. Army Reserve, was the superb operations chief for civic action. He held together the day-to-day operation for a $138 million program executed by seven U.S. Brigade Combat Teams and a U.S. and Iraqi combined force of 62,000.

e. Major Douglas Davids, U.S. Army, was the main strategist behind the division's civic-action program. He is a leading expert in counterinsurgency campaigning and works at the doctorate level. Always have a Doug Davids at your side.

f. Lieutenant Colonels Kevin Dixon, John Cross, Matt Elledge, Joe Gandara, Troy Smith, Chris Hall, and Chris Pease were combat leaders who could surmount the impossible. They never wavered in their mission and rose to unmatched brilliance during the heat of the Battle of Baghdad.

g. The S9s (Civil-Military Operations officers) were the quiet professionals who brought victory.

LTC Rick Summers	*172nd Stryker Brigade Combat Team (SBCT*)
MAJ Herb Joliat	*1st BCT/4th Infantry Division (1/4 ID)*
MAJ Jeff Smith	*2nd BCT/4th Infantry Division (2/4 ID*)
MAJ Jose Acosta	*4th BCT/4th Infantry Division (4/4 ID*)
MAJ Paul Schmidt	*2nd BCT/101st Airborne [Air Assault] Division (2/101 AASLT)*
MAJ Bren Workman MAJ John Sottnick	*506th Regimental Combat Team/101st Airborne Division (506 RCT/101 AASLT*)
MAJ Roland Bennett	*1st BCT/10th Mountain Division (1/10 MTN)*
MAJ Philipe Moon	*2nd BCT/10th Mountain Division (2/10 MTN)*
MAJ John Culberson	2*nd Brigade, 1st Armor Division (2/1 AD*)

h. Lieutenant Colonels Marty Nelson, 448th Civil Affairs Battalion and Bill Woods, 414th Civil Affairs Battalion, the Army's finest Civil Affairs commanders in the force today. These two officers and their command Sergeants Major lead their great formations from the front, out on point every day, driving their Civil Affairs teams to accomplish the unachievable.

Lieutenant Colonels Diana Brown and Neil Aihle took over Operation Holland Light after the death of Lieutenant Colonel Daniel Holland who was killed by an improvised explosive device (IED). The narrative of Operation Holland Light herein is based on their report.

The personal commitment, creativity, and astuteness of these officers formed the backbone that won the day. They are the greatest collection of professionals you could ever go to war with and never ceased to impress me with their achievements. These were the soldiers who fought the Battle of Baghdad block by block, these quiet few.

The G9 and brigade S9s, the quiet few who brought success to the Battle of Baghdad, 10 November 2006

***

Photographs are from the Assistant Chief of Staff, G9, Public Affairs Office, and members of the 4th Infantry Division (Mechanized).

***

1. Introduction

Since the September 11, 2001 attacks on the World Trade Center, the U.S. Army has moved through its process to retool itself to fight an insurgency. This feat has been on par with transforming the Army prior to and during World War II. From 1938 to 1945, the Army went from flat-brimmed helmets and wool uniforms designed for trench warfare to a uniform designed for maneuver warfare. We went from biplanes to an Army Air Force with pursuit aircraft, long-range bombers, and cargo planes. New rifles, tanks, artillery, and fielding an army of 16 million all occurred during this time. Doctrine has changed from an expeditionary force using cavalry, infantry, and artillery to an army that used combined arms action to defeat its adversaries and then, after liberating enemy territory, assumed occupational duties and exerted military governance.

With the Global War on Terrorism, the Army has transformed to defeat the enemy. The enemy of today, however, is far different than the enemy faced during World War II. As such, the focus now is on small unit operations vice the Army-wide operations conducted in Europe, such as the U.S. Third Army's fight during the Battle of the Bulge. Today, we are transitioning from combined-arms action to counterinsurgency (COIN) operations.

In understanding the new way of war, we have changed our doctrine and articulated several approaches or methods to conduct COIN operations. The Army Counterinsurgency Field Manual (FM) 3-24 offers several methodologies such as clear, hold, build, combined action, and limited support. These methods can be used together or unilaterally, but they are not the only methods. In actuality, COIN operations are a combination of strike operations to kill or capture the insurgent *and* civic action to remove or reduce the causes of instability. Simple enough, right? Well, that is the formula, and whether the insurgency is defeated or not depends on the balance of the application of a COIN formula: COIN operations = strike operations + civic action.

COIN operations = strike operations + civic action.

A COIN campaign or operation is one of the most complex military engagements to be executed. The complexities of the operational environment cause confusion. Factors such as the societal structure, culture, language, religion, geography, levels of corruption, competency of the host nation government, the extent of the rule of law, whose law, the will of the American people to endure the fight, and hundreds of other undertows make a seemingly impossible situation hopeless. Hence the complexities must be simplified, so the sergeant leading the patrol and the soldier at the point will think before they act and know when to kill and when to build. Codifying COIN operations as a mix of strike operations and civic action reduces the ambiguities that the sergeant and private face every day. If COIN operations are not simplified, the officer and noncommissioned officer can become unfocused, and subsequently, mistakes are made. In the COIN fight, a tactical error can bring strategic failure.

This monograph examines the Battle of Baghdad as a case study for civic action with the intent of contributing to the writings on this important topic. The description of operations conducted during this fight serve as a playbook for creating a favorable environment for current or anticipated operations. Also, these kinds of operations can gain the support, loyalty, and respect of the people for their local and national government. This gain can be achieved by assisting communities in conducting health, welfare, and public works projects; improving the economic base of the nation; and improving the standing of the host country and allied forces with the population.

The details of conducting a civic-action program at the divisional and Brigade Combat Team levels of command are presented herein. This monograph looks at the Battle of Baghdad during 2006 and the civic-action program that provided for the societal underpinnings of 11 million people in four Iraqi provinces, consisting of 16,664 square miles. This area included the capital city of Baghdad, the center of gravity of the campaign in Iraq.

The name “block by block” came from a variety of sources:

a. Epitomizes the strategy for civic action in an urban environment or city by focusing efforts on the neighborhoods, which are the core foundation of a city society and the lowest organized congregation of the populace.

b. Reflects on the Iraqi government’s desire to rebuild Iraq “block by block,” which refers to building construction using masonry blocks or bricks.

c. Delineates how to secure a city using a city block as the geographic reference. We battle-tracked civic action by using the *muhallahs* or neighborhood blocks of the City of Baghdad.

Chapter 2 addresses the operational environment that challenged civic actions of the Multi-National Division, Baghdad (MND-B). The Baghdad area was the "Red Zone" where every detail of planning and every operational move was critical. In any event, a focused enemy used to his or their advantage the economic, cultural, and historic underpinnings of the struggle.

Figure 1. The Red Zone, operational environment in the MND-B area of responsibility

Figure 2. Sadr City, 22 July 2006, indicative of Baghdad's infrastructure problems, an intertwined mix of haphazard solutions

2. The Red Zone

The overarching mistake made during the course of the Global War on Terrorism has been misunderstanding the operational environment. During the Cold War, we could template a Combined Arms Army, identify where command and control nodes were in Soviet formations, and define completely the enemy order of battle. We studied military history and took staff rides to old battlefields to see how the ground was fought and won during the American Civil War, World War I, and World War II. We took courses in the native language of the country we were operating in and moved around these forward areas, not as a tourist, but as a resident. We took the time and applied the effort to understand how our enemy was organized, how he fought; we learned the ground as our own and lived the history behind the next war.

We have not done that well for the Global War on Terrorism, and what we continually fail to understand is our enemy. We say that the enemy is not organized, yet he issues orders to his soldiers daily. We believe no command structure exists, yet the grand emirs, emirs, and brigade commanders are in charge and directing their forces. We believe the enemy is an undisciplined force, but he wears a standard uniform, has an area command that provides active security for the populace, and is proud to be a Mujahideen.[1] In addition, we believe that he has no set of guidelines to fight by, yet he applies a 1,400-year-old jihad doctrine[2] and uses our doctrine against us.

Part of predeployment planning was realizing the necessity of understanding the operational environment—for example:

a. Iraqi society, Arabic, local customs, geography, and history of a land 5,000-years old

b. Construction of the enemy forces for determining their goals and motivations

c. Inner workings of the society for constructing a program to move the populace away from the insurgency.

We had to bury American arrogance and learn Iraqi pride. That led to finding ourselves in the Red Zone, which in American terms is best defined in the novel *The Red Zone*:

> In football, once the offense crosses the other team's twenty, it is called the red zone because every move, every penalty,

> every inch, every detail is critical. In a game of violence, nowhere is the primal urge to attack and defend more clear. Here is where the lust for blood boils most clearly to the surface. Here is where the winners are separated from the losers, the weak from the strong. Mistakes in the red zone are known to cost people their jobs, their careers, and therefore their lives.[3]

In Baghdad terms, the Red Zone is the unsecured areas of the city. The residents tell us the Red Zone is everything outside the Green Zone, which is the nickname of the old Presidential Palace area of Baghdad, where Saddam Hussein had his government offices. The Green Zone is a fortified complex where the coalition houses its political and military headquarters, as well as where the Government of Iraq is located.

The Red Zone is where the Battle of Baghdad was waged. An area critical to the war, it is land we would attack and defend and where mistakes meant lives. Therefore before entering the Red Zone, we first understood the history behind COIN operations, then the area of responsibility, the enemy order of battle, what the enemy was fighting for, and what the citizens of Baghdad wanted. That meant analyzing the operational environment by narrowing down the strategic view, through an operational "lens" to determine the tactical outcomes, and all from the enemy's viewpoint.

The History Behind the Insurgency

Studying the history of 20th century insurgencies and of Baghdad led to insights that guided planning.

The average insurgency during the course of the 20th century took 14 years from start to finish. Time was on the side of the insurgent, and the center of gravity for the Battle of Baghdad was the American people's will to sustain the fight. As the 4th Infantry Division entered Baghdad, the Global War on Terrorism was in its fifth year, moving past the duration of America's involvement in World War II. This revealing statistic strains America's patience and public support for the war. Hence, the ultimate factor in fighting a modern day insurgency is managing perceptions and expectations, for maintaining the public's will to endure the fight, while the elements of national power forge ahead to achieve the defeat of the insurgency. Time is on the side of the insurgent.

Looking back over the 20th century, we learned that insurgencies are defeated by achieving victory at all levels of war and by defeating both the insurgent and the insurgency. This methodology focuses on winning at the strategic, operational, and tactical levels of war:

a. At the *strategic* level, COIN campaigns achieved victory when all elements of national power were applied. Codified today as *diplomatic*, *informational*, *military*, and *economic* elements, when applied in supporting and supported roles, the insurgency is defeated. Historically, the best way to strategically defeat an insurgency is using diplomatic efforts to isolate external support to the insurgency, information campaigns to discredit the insurgency's goals, military force to protect the populace, and economic development to increase the populace's quality of life. When the elements were not applied, success in defeating the insurgency was fleeting.

b. At the *operational* level, in order to defeat the insurgency, the causes of instability had to be removed or reduced. Whatever made the populace move to support the insurgency must be identified and addressed. The overall goal is to remove the support base from the insurgency. In general, 10 percent of the populace comprise the guerrilla force, 5 percent are the antigovernment subversives, leaving the balance of 85 percent available to support the insurgency. Historically, the best way of defeating an insurgency at the operational level of war is to win the support of the 85 percent population base by giving them what the insurgency is promising. Simply put, out-compete the insurgency by giving the people a better quality of life and solving their collective problems.

c. At the *tactical* level of war, in order to defeat the insurgency, defeat the insurgent. This meant focusing on the enemy's order of battle and systematically attacking the components by using lethal and nonlethal means. Historically, the best way of tactically defeating the insurgent was cutting his supply lines or "buying out" his logistics base.

Ian Frazier provided the historic importance of Baghdad in his article "Invaders, Destroying Baghdad." He gives the Arab and Persian names for the city:[4]

a. Madinat as-Salaam (City of Peace) is Arabic.

b. Bagh (God) and Dad (given) is Persian.

Baghdad was once the crossroads of the world and was known as all that was once great with Islam. Baghdad was revered as the birthplace of civilization and was the epicenter of wealth, power, religion, and the Islamic empire. Within 40 years after its founding in 762, Baghdad had achieved greatness that would perpetuate itself to modern times. Named by its founder Jaffar Al Mansour (member of the Abbassid line of caliphs) as Madinat as-Salaam, Baghdad became a seat of poetry, education, and knowledge. During Baghdad's height, it was legended that almost all of the residents could read and write. Checkers, chess, and backgammon were commonly played. Greek was translated into Arabic. The city was filled with libraries, parks, bathhouses, and palaces. In fact, the palaces housed the caliphs and were constructed of marble, rare woods, jade, and alabaster with carpets and wall hangings by the thousand. Its grounds had fountains and interior gardens. Servants sprinkled guests with rosewater and powdered musk and ambergris. Arts and sciences flourished—literature, music, calligraphy, philosophy, mathematics, chemistry, history.

All that was once great about the City of Peace came to an end by the hand of Hulagu Khan beginning 29 January 1258. A grandson of Genghis Khan, Hulagu concluded his approach on Baghdad from the east and began his investment on the outskirts of Baghdad. His preparations began with a bombardment, and by 4 February the outer wall was breached. On 10 February, Baghdad surrendered, after the Caliph Al Musta'sim came out of the city and surrendered. The Mongols swept into the city on 13 February, which began a week of massacre, looting, and fire; the Mongols killed almost everybody. For 7 days, the Mongols dominated the city by sacking it, killing an estimated 1 million and plundering its gold, jewels, and silver. The caliph was imprisoned, and later killed, and the city was occupied by the Mongols. The Mongol occupation force was approximately 3,000 and was tasked to rebuild the city. Little was accomplished; the city disintegrated with disease and famine, and decades later the city still laid in ruin, with some aspects of the infrastructure only recovering in the 20th century. Today the citizens of Baghdad view the American liberation in 2003 as another sacking of Baghdad.

A final lesson was that we did not follow historical siege typology when taking the City of Baghdad. The siege is typically organized into seven phases:[5]

a. Approach—lining up for the attack
b. Investment—laying in the siege, the encirclement
c. Preparation—timing and method of determining the assault
d. Assault—the walls are breached and defenses penetrated
e. Dog Fight—the break through, looting, plundering, and wanton violence
f. Domination and Occupation—the besiegers take control of the city
g. Withdrawal—when the besiegers depart, voluntarily or by force.

Analysis had shown that all we had done to date was the Approach—skipped the Investment and the Preparation and went right to the Assault. The defending units in Baghdad had taken the opportunity to flee and go to ground as the coalition moved from the Approach to the Assault and did not complete an encirclement of Baghdad, which would have contained the defending forces and limited the opportunity for the disbanded Iraqi Army to turn into a guerrilla force. We were actually still in the siege, and fighting a Dog Fight to gain Domination and Occupation of the city.

From the history of insurgencies and of Baghdad, the staff gained the following insights:

a. The typical 14 years from start to finish of an insurgency put time on the side of the insurgent. A way to use time against the insurgent is to complete civic-action projects within 120 days and that address the immediate needs of the people.
b. The diplomatic, informational, military, and economic formula was not working. Though the military element of national power was in full play, the Department of State had not assumed responsibility for the governance and economic lines of operation. Concurrently, the lack of a cohesive U.S. government information campaign to support U.S. action in

Iraq created opportunity for enemy propaganda campaigns to advance unchecked.

c. Baghdad had been a sacred city to its residents, and all they wanted was to make it great again. Hence the focus was civic action not only on immediate needs but a revitalization of the city.

d. We were still in a siege. In fact, the enemy had laid in a counter siege and any civic actions were vulnerable to attack, as part of the ongoing Dog Fight.

Understanding the history behind an insurgency and gaining historic appreciation of battleground led to a focus on the enemy order of battle. Though the enemy is organized with grand emirs, emirs, area commanders, vanguard forces, foot soldiers and supporters, a generic appreciation of the enemy structure included a sufficient knowledge of the enemy order of battle. Also, understanding the enemy's organization would lead to understanding their operations, and subsequently how to defeat the elements of the insurgency.

The Enemy Order of Battle

In its simplest terms, the enemy in Iraq is codified as the insurgent and insurgency. By definition, an *insurgency* is an organized movement aimed at the overthrow of a constituted government through use of subversion and armed conflict (Joint Publication 1-02, DoD Dictionary of Military Terms). Members of the insurgent force are organized along political lines to support political, economic, social, military, psychological, and covert operations. As also covered in the 2006 monograph, *One Valley at a Time*, the scenario is divided into three parts: the guerrilla, underground, and auxiliary.[6]

The Guerrilla. In JP 1-02, *guerrilla* is the combat element of the insurgency. *Guerrilla warfare* is the military and paramilitary operations conducted in enemy-held or hostile territory by irregular, predominantly indigenous forces.

In U.S. Army doctrine, the guerrilla is the overt combat element of the insurgency. The members of the guerrilla force organize under military concepts to conduct military and paramilitary operations. See Table 1. Their duties usually include all the overt actions of the insurgent organization but may include covert and clandestine operations. They are usually most active in insurgent-controlled or contested areas. However, when the insurgency calls for operations in government-controlled areas, the guerrilla may conduct these operations.

Table 1. The Enemy Defined: Guerilla Forces

Level	Organization/Location
Global/strategic	Front commander/grand emir Iraq-wide
Strategic/operational	Subfront commander/emir Baghdad
Operational/tactical	Area commander Kararhi I, II and Rusafa I, II
Tactical	District commander beladiyahs (boroughs)
	Cell leaders hayys (wards)
	Foreign fighters: revolutionaries or vanguard city-wide
	Indigenous (homegrown) fighters muhallahs (neighborhoods)
	Suicide vehicle-born improvised explosive device (SVBIED)
	Vehicle-born improvised explosive device (VBIED)
	Improvised explosive device (IED)
	People's Revolutionary Army (ERP)
	Suicide bombers
	Surface-to-air missile (SAM) engagers
	Indigenous militias specific locations

The guerrilla force tries to gain support for the insurgency through propaganda, coercion, and terror. Terrorism is a "weapon system" or technique that the guerrilla applies to achieve his goal. If he cannot gain active support, he will seek *passive* support.

A situation in which a population has become submissive to the guerrilla or neutral, siding neither with the insurgency or the government, is an unacceptable state to COIN operations. A people's neutrality is dangerous, and the silence on the part of the populace concerning insurgent activities is considered passive support for the insurgent.

As U.S. Army doctrine noted the guerrilla has advantages, doctrine also states that he has vulnerabilities. The guerrilla normally lacks the personnel and the logistics to intentionally become decisively engaged with COIN guerrilla forces. In many circumstances, as found in post-World War II doctrine, the proportionality of the COIN guerrilla campaign will force the guerrilla's potential recruiting base to diminish. Effective COIN guerrilla operations will not regenerate follow-on guerrilla insurgent forces. Hence, the guerrilla's vulnerability, open to exploitation, is his source of supply and recruitment. Convince the potential recruit not to join the guerrilla, protect the shopkeeper so he is not forced into supplying the guerrilla, inflict combat losses, force the guerrilla deep into untenable sanctuary, or interdict their supply lines leads to degrading the guerrilla's ability to sustain.

Additionally, exploit the guerrilla's fear of capture. This vulnerability is paramount to a successful COIN guerrilla campaign. First, disseminate, through multiple means and sources, the outcome of capture. Reinforce that the guerrilla is a criminal; internment will mean long (unknown) separation from his family—that is, if he is captured, the family will be disgraced publicly and all which the guerrilla is fighting for will be lost. Any retribution by the guerrilla force against the family of a captured guerrilla is also a key vulnerability and subject to exploitation.

The guerrilla's operational weaknesses often includes security, which requires extensive resources and slows down his responsiveness; bases that are difficult to acquire and operate; and the lack of technology or the ability to maintain captured high-technology items. Another operational weakness may be the lack of sophisticated communications, which could require the guerrilla to spend an excessive amount of time preparing to launch an operation. The guerrilla's dependence on popular support suggests a potential vulnerability because if that support wavers or is withdrawn, the guerrilla will not be able to operate effectively.

The Underground. In Army Regulation (AR) 310-25, Dictionary of U.S. Army Terms, the *underground* of an insurgency is defined as combat support of the insurgency and a covert unconventional warfare organization established to operate in areas denied to the guerrilla forces or conduct operations not suitable for guerrilla forces.

The underground organizes into compartmented cells. Cells are formed within various political subdivisions, sectors, or areas, such as the U.S. equivalents of counties, towns, and neighborhoods. The underground environment may be urban or rural. Underground members commit sabotage, intelligence gathering, and acts of deception. They are the political and financial wing of the insurgency and are the main element focused on subverting the government. The underground supports the guerrilla and the auxiliary at the direction of the area commander or resistance front commander. They operate in small cells and are linked to other cells through key leaders, intermediaries, or facilitators. The underground operates covertly or overtly, uses safe houses to hide insurgents, and will move people and supplies along routes and in vehicles to avoid detection.

The Auxiliary. In AR 310-25, the *auxiliary* is the organized civilian support to the tactical action arm of the insurgents. In unconventional warfare (UW), it is that element of the resistance force established to provide the organized civilian support of the resistance movement.

The success or failure of the guerrilla force depends on its ability to maintain logistic and intelligence support. The auxiliary fills support functions by organizing civilians and conducting coordinated support efforts. The assistance of the civilian population is critical to the success of the resistance movement; it provides security, intelligence, and logistic support for the guerrilla force by using civilian supporters of the resistance. The auxiliary conducts clandestine support functions by organizing people on a regional, district, or sector basis depending on the guerrilla force formation. To sustain guerrilla operations, the guerrilla force has numerous needs:

a. Movement support
b. Acquisition of supplies
c. Operational information and intelligence
d. Medical facilities
e. Counterintelligence to establish an outer zone security early warning network to detect, if not deter, enemy penetration attempts
f. Recruiting support
g. Compartmented communication systems
h. Current information on terrain, weather, civilians, and local resources
i. Deception operations support
j. Manufacture and maintenance of equipment.

Table 2 delineates the composition of the subversive (underground) and logistics (auxiliary) elements of the enemy order of battle.

Table 2. The Enemy Defined: the Underground and the Auxiliary

Level	Organization/Location
Operational	**Underground**
	Political leaders
	Ideologues
	Financiers/profiteers
Operational/ tactical	Facilitators/black marketeers
	Recruits
Tactical	IED makers
	ERP suppliers
	SAM suppliers
	Small arms
Tactical	**Auxiliary**
	Intel/early warning
	Subsistence
	Safe houses
	Transporters
	Caches
	Electromagnetic lines of communication (LOCs)
	Ground LOCs

With an understanding of how the enemy was constructed, the next step was to determine what the enemy wanted and why. What were the enemy's goals and motivations? Why did they care that American forces were in Iraq? And why did they now—after the coalition had liberated the country from Saddam's tyranny—actively oppose the coalition's efforts to rebuild the government, repair the infrastructure, and bring Iraq forward as a good neighbor to the world community?

The Enemy's Goals and Objectives

The predeployment analysis concluded that the enemy sees Baghdad as its center of gravity and Baghdad is what they are fighting for. The global insurgency is led by Al Qaeda and organized into fronts of resistance that are active in the U.S., Europe, Russia, Chechnya,

Afghanistan, Turkey, Lebanon, Palestine, Iraq, the Arabian Peninsula, East Africa, Kashmir, Australia, and the Southeast Pacific.

Al Qaeda was in the Red Zone with us. The enemy was also in transition from a guerrilla war to a war of movement as they refined their organization, built combat power, developed new tactics, attacked in organized groups, and equipped and uniformed their guerrilla forces into armed militias or as they called themselves—armies (jaysh in Arabic).

But why? Why did Al Qaeda see so much importance in Baghdad? What do they want? The answer is well documented in the speeches and messages from Osama bin Laden, the leader of Al Qaeda:

> *12 November 2002*
> [U.S. Vice-President] Cheney and [U.S. Secretary of State] Powell killed and destroyed in Baghdad more than Hulagu of the Mongols.
>
> *18 October 2003*
> In conclusion, I say to the American people we will continue to fight you and continue to conduct martyrdom operations inside and outside the United States until you depart from your oppressive course and abandon your follies and rein in your fools. Baghdad, the seat of the caliphate, will not fall to you, God willing, and we will fight you as long as we carry our guns. And if we fall, our sons will take our place.
>
> *29 October 2004*
> We gained experience in guerrilla warfare and in conducting a war of attrition in our fight with the iniquitous, great power—that is, when we conducted a war of attrition against Russia with jihad fighters for 10 years until they went bankrupt, with Allah's grace; as a result, they were forced to withdraw in defeat, all praise and thanks to Allah. We are continuing in the same policy—to make America bleed profusely to the point of bankruptcy, Allah willing. And that is not too difficult for Allah.
>
> *16 December 2004*
> The most important and serious issue today for the whole world is this Third World War, which the Crusader-Zionist coalition began against the Islamic nation. It is raging in the

> land of the two rivers. The world's millstone and pillar is in Baghdad, the capital of the caliphate.

Bin Laden states that the focus on Baghdad is for three reasons. First Baghdad and Iraq is the place where Al Qaeda can destroy the U.S. and purge the western influence led by America from the Muslim world. Purging the western influence from the Middle East will allow for the protection of the Muslim religion. Secondly, the focus on Baghdad is for Baghdad's defense. Baghdad is the capital of the Islamic Empire; it was sacked by the Mongols, sacked again by the Americans, and it is Al Qaeda's responsibility to defend the capital, in order to reinstate it as the capital of the caliphate that Al Qaeda is charged to resurrect. Finally, Al Qaeda wants to reveal to the Islamic world the holocaust that the west has brought to one of Islam's treasured cities.

The assessment was that Al Qaeda sees Baghdad's occupation by the coalition (Domination and Occupation phase of Siege Typology) as a crusader siege of Baghdad. Al Qaeda is applying classic guerrilla tactics to break the coalition siege by encircling the encirclers or laying siege to the besiegers and have enlisted the active support of other Muslims in this fight. Although Al Qaeda has aligned for short-term gain with a variety of anticoalition forces, a greater issue exists below the surface: the internal dynamics of the Iraqi society.

Muslims are a people steeped in their past. The wars and battles in their recent collective memories are those of the Crusades. Iraq, however, is a country marred by its present, and that present involves the British mandate.[7] Underneath it all, what plagues Iraq each day is the constant conflict of jockeying between Arab and Persian, being pro-western or fundamentalist, and the struggle to industrialize or keep the land between two rivers agrarian.

The story of the conflict in Iraq is based on thousands of years of foreign conquest, domination, and internal struggle in an area encompassing the Fertile Crescent. It is a story of a desert people moving to the east, meeting with another culture moving from the west and converging along the Euphrates River. It is complicated by their common belief in Allah but opposed by their divergent interpretation of Mohammed's succession. More so it is about economics, specifically oil production and water distribution.

Hence, the struggle for control is on to achieve resolution of Iraq's internal struggle. Should Iraq move to a confederation of the Kurdish, Sunni, and Shi'a states and transition from a semi-industrial/agricultural society to a 21st century industrial culture integrated into the world's economy, based upon an economic stance as a power and water provider for the Middle East and Iran? Should Iraq become an Islamic Republic under Sharia law? Should Iraq retain its central government with provinces aligned underneath it, or should Iraq become a western democracy? Much more is needed to be considered.

What Does the Citizen of Baghdad Want?

Studying the Red Zone surfaced that the coalition forces did not fully understand the operational environment in Baghdad. The focus needed to be on understanding the Iraqi society, the language, and the culture, local geography such as street names, local customs, and how the Baghdad citizen saw coalition forces. The goal was to identify the daily abrasion to the very populace the forces were there to protect.

Robert Burns said it best: "O' would some power the giftie give us, to see ourselves as others see us"[8]—insight applicable to the Battle for Baghdad. The best means found to understand how the average Baghdad citizen saw the coalition's campaign in Iraq was from basic atmospherics. The experience of citizens on the streets around Baghdad can be very revealing.

> When the Americans fire back, they don't hit the people who are attacking them, only the civilians, said Osama Ali, a 24-year-old Iraqi who witnessed the attack [in Baghdad]. This is why Iraqis hate the Americans so much. This is why we love the mujahideen. New York Times, 14 September 2005.

> Baghdad—At dawn, Leith Hassan went to pick up bread for the family's breakfast. Jumping on his red bicycle, the 14-year-old whizzed down the street, passing the carpet store at the corner. Just then, a driver detonated a bomb hidden in his car. One red-hot piece of shrapnel severed the boy's left leg. Later, from his hospital bed, Leith tried to reassure his father. 'Don't worry, Daddy,' he said. 'I'll be OK.' Leith,

> who had dreamed of becoming a pilot, died that afternoon. "Living Under Constant Siege in Baghdad," LA Times, 28 September 2005.

> Every Muslim should know that his life is not his own; it is the property of this violated nation for whom men have shed their blood. No other issue should take precedence over work for the Umma [community of the faithful within Islam]; it is an obligation for every Muslim.[9]

A summary of the perceptions, experience and expectations of the citizens of Baghdad to the American invasion into Iraq can be found in a *Washington Post* article, a portion provided here:

> In the chaotic, hopeful April of 2003, Baghdad's Karrada district was one of those neighborhoods where residents showered flowers on U.S. forces entering the capital ... more than 2 years after flowers cascaded onto the arriving Americans, what's being thrown on Karrada's streets has changed as well. Mohammed and the neighborhood watched as American soldiers on patrol grew irritated at an Iraqi who had left his car in the street to run inside a store on an errand, blocking their armored convoy. The Americans took one of the empty plastic water bottles they use to relieve themselves ... when the Iraqi driver ran out to move his car, an annoyed American plunked him with the newly filled bottle and rolled on, Mohammed said. 'Now in Iraq, every morning you kiss them goodbye,' Jassim Mohammed (age 41) said, referring to his wife and children, 'because you don't know if you will be back or not. Everyone in Iraq does that now.'[10]

Even enemy recruiting films such as the video clip Dirty-Kuffar,[11] by Sheikh Terra and the Soul Salah Crew, displayed America as a Nazi infidel pillaging the world and corrupting innocence.[12]

The MND-B public opinion poll of September 2005 identified the public's concerns and delineated how the citizens of Baghdad perceived coalition efforts to build a democracy in Iraq:

a. Top three priorities were electricity, security against terrorism, and water.

b. Seventy-nine percent had no confidence in the multinational forces to improve the situation in Baghdad.

c. Eighty percent reported less than 5 hours electricity each day.
d. Seventy-eight percent had confidence in the Iraqi Army and 79 in the Iraqi Police.
e. Sixty-five percent were dissatisfied with the local government.
f. Popular support for the insurgency was steadily declining; only 25 percent expressed confidence in the armed national opposition to improve the situation in Baghdad.
g. In the prior 6 months, 20 to 23 percent were unemployed.

In the same time frame, the poll was reinforced in September 2005 interviews of Iraqi military officers in Baghdad; they provided insight on the causes of instability and suggested how to reverse or reduce them. These candid observations were collected during a series of unofficial conversations with officers from the Marine Corps Center for Lessons Learned and the Center for Army Lessons Learned mobile training team (MTT) supporting the NATO Training Mission-Iraq. The statements of the assembled group were related as close to verbatim as possible. The method of recording was via notes written during the conversations, and the opinions and recommendations presented are those of the Iraqi officers only.

Eight of the 11 officers in the assembled group held master's degrees. The composition of the group was one major general (retired), three colonels, four lieutenant colonels, two majors, and one captain. The MTT officers and the Iraqi officers had been working together daily for over 3 weeks. In the words of one of the Iraqi officers, "we have opened our hearts to you." Some of the measurable claims have been verified with other agencies working outside the "wire." Table 3 provides a summary of discussion on six topics.

Table 3. Result of Topics Discussion

Topic	Result
Political issues	Return all employees to previous jobs. [Restated during infrastructure conversations.] Dissolve all fundamentalist or religious parties. Do not have or allow any religious political parties. [Firmly believe in the separation of religion and state.] Remove the outside Iraqis from positions of authority. The ones from outside are only here for the money. They are here for personal issues and not for the country. These outsiders will leave once they make their money. Disband the militias so people are not afraid. Every party has its own militia, and they cause problems in elections. If I vote against them, they will kill me. Iraqis are at personal risk to remove 'bad' people through the election process. In recent election, they [religious leaders] urged us to vote for certain purposes/people. If we did not vote this way, we are not religious. They are using religion to control us and our vote. We vote in accordance with the desires of our religious leader. Iraqis do not trust the election process or the people running the elections.
Election and the voting process	The ballot requires pictures; many Iraqis are unable to read. The election official then stands with them to help them vote, and they tell them for whom to vote. Remove the men at the voting places and put in women. Women in our society must follow the rules, and men are allowed to modify the rules. Make the women swear an oath to be fair. There will be a higher level of confidence. There must be an international organization that specializes in monitoring the fairness of votes—an organization that has done it for other countries. It can be multinational. Arabs can participate if they are really a part of the organization and have worked on other elections. During the voting (last election), for 3 days, there is curfew. It was 10 kilometers from my house to my voting place—too far to walk. In the last 2 hours of the last election they moved the voting box, but many people could not vote. The person who controls the registration list knows who has been to vote and who has not. The reason there was such a large turnout was because when someone did not come to vote, their ballot was filled in for them.

Table 3. Result of Topics Discussion (continued)

Topic	Result
Election and the voting process (cont.)	The main problem is the person who manages the ballot boxes. They have relationships with different organizations. The ballot-box minder will decide the vote. Happened in my neighborhood. I watched an old lady take a ballot, and she could not read or write. She told the man she wanted to vote for Allawi, and I saw the minder select someone else [chose name from own party]. It is not fair that my vote can be given to someone who I did not vote for. This makes it too easy to change what the people want. The system of voting for a list of people supports the religious influence. There should not be a religious list—just individuals only. Resentment exists towards the ex-pats who came back into leadership positions. The general feeling is that 'they were dancing in the discos of Europe for 10 years while we were here suffering.' They are not here for the Iraqi people; they are here to make their money and leave.
Infrastructure and services	Gasoline is a 4-hour wait, and when I get home there is no power or water. Sometimes I go 3 days without water. My family washes their clothes in the river because it is there [Baghdad]. Destruction in 1991 was much larger, but Saddam was able to rebuild because the people who ran the infrastructure were still in place. I do not see any new buildings for Iraqis. All we see is concrete walls all over our cities. Provide us with something as simple as paving the roads. The U.S. should push the Iraqi government to improve services.
Insurgency	Ambassador Bremer disbanding the Iraqi Army and government was a bad thing. Now the salaries [monthly allotment] paid to these former Army and employees is not enough, they can receive more money from the insurgency. They have to feed their families. Question—If Ambassador Bremer had not disbanded the military, would there be this insurgency/secular tension? Answer—No problem; Iraq would have a different image.

Table 3. Result of Topics Discussion (continued)

Topic	Result
Insurgency (cont.)	Parties [insurgent groups] are serving outside countries [Iran and Syria]. They do not want Iraq to have settlement. If Iraq gets settled, they think it will be their turn. They have money, authority, and power [from their countries]. Neighboring states [Iran and Syria] want us like this … they do not want to see us regain power. Many Iraqis believe that the U.S. does not want a settled Iraq because they want to move the terrorism fight from the U.S. to Iraq. We see this in your news. Most of Iraqi people are threatened by the militias. Our children want peace.
Education	Educated people believe in a more liberal system of government. Uneducated Iraqis are a simple people who follow traditional ideas—religious leaders tell them who to elect and follow. U.S. must educate the Iraqi people. The more we educate the people the less problems we will have. Before Saddam we used to have people go to university abroad. That has stopped. We need more people to be educated. [This issue was narrowed to before Saddam's relationship with the western world deteriorated. The point is that they want to regenerate the overseas education that once existed.] My brother's son needs special medical attention. There are no special doctors or even teachers. They are either dead or have left the country out of fear. [Repeated during infrastructure discussions.] We need education; our schools and universities have been destroyed.

Table 3. Result of Topics Discussion (continued)

Topic	Result
Military service	I am a member of the new Iraqi Army and cannot be proud. I must hide my uniform. I wear civilian clothes to work. When my small child asks me where I work, I tell him I am a school teacher.
	If I go away for 4 days and my neighbors notice, I will become a target.
	In north and south Iraq, there is not as much problem with being in the military. Problem is in the triangle. In southern Iraq, they [members of the Iraqi military] can wear uniforms and walk along the street. In Baghdad, I wear civilian clothes and have to hide my job even from my relatives.
	Mix the new Iraqi Army with the old army to provide us with order in the future. They [old Army] will be welcomed in the new Army.
	We should retire the other [more controversial] groups and pay them their pensions or give them civilian jobs [i.e., remove incentive to serve insurgency].
	If a former officer wants to join the new Iraqi Army, they examine his record; if he is a former Ba'ath Party member, he is not allowed to join new Iraqi Army... let the Ba'ath Party members return.

Assessing these various issues and insights expanded our appreciation of the operational environment in and around Baghdad. This understanding was essential as we sought to apply the theories of civic action to the January 2006 situation in Baghdad.

3. The Enemy We Fought

The enemy we fought in Iraq in 2006 was ubiquitous and came from many directions. The enemy came from several groups who are diverse in their beliefs, held varied agendas, and held a myriad of objectives.

In some cases, the enemy we faced was the *true believer.* He believes in his religion, his tribe, his family, even his country, but what he truly believes is that his cause is just. He is so firm in his conviction that he has taken up arms to defend it. His beliefs are absolute, and the jihad he wages is for an eternity.

> ***The True Believer***
>
> *Somewhere a true believer is training to kill you.*
>
> *He is training with minimum food or water, in austere conditions, day and night. The only thing clean on him is his weapon. He doesn't worry about what workout to do—his rucksack weighs what it weighs, and he runs until the enemy stops chasing him.*
>
> *The true believer doesn't care 'how hard it is'; he knows he either wins or he dies.*
>
> *He doesn't go home at 1700; he is home. He knows only the Cause.*
>
> *Now, who wants to quit?*
>
> *— Unknown source*
> *at Fort Bragg, North Carolina*

In many other instances, the enemy we fought was the end-state. The variety of actors on Iraq's stage made the MND-B area of responsibility a highly complex and very lethal operational environment. Throughout the Baghdad, Babil, Karbala, and Najaf provinces, the MND-B fought Sunni extremists, Shi'a radicals, Iranian subversion, Syrian interference, and Al Qaeda. Groups worked together, against each other, in mangled alliances, in separation, or a combination thereof, but over the course of time, each entity emerged as a distinct threat that challenged the end-state and complicated the operational environment. As the year ended, the enemy we fought was the end-state—a conflict to determine the final outcome or objective(s) of coalition efforts, the newly elected Government of Iraq, the people of Iraq, the competing external factors from Saudi Arabia, Turkey, and Syria.

In 2006, as the Battle of Baghdad raged, the enemy manifestation morphed over time. At first the enemy was focused against the U.S., then at the end of 2006, each other.

The War Against the Occupier. The operational environment that existed as the 3rd Infantry Division transitioned authority for the MND-B to the 4th Infantry Division on 7 January 2006 was with the populace viewing the U.S. as the occupational force. The enemy wanted coalition forces removed. The predominant target of enemy attacks was against U.S. forces to make losses in personnel and equipment great enough to force a withdrawal. All factions focused on targeting coalition convoys, routes, and patrols. In the mix were secondary attacks against the populace, primarily of Shi'a pockets to force an escalation of violence and tumble Iraq into civil war or anarchy. This effort would destabilize the nation, force a greater expenditure by the U.S., and bleed the coalition into bankruptcy.

At the end of Ramadan in 2005, both Shi'a and Sunni clerics made public prayers and speeches acknowledging the breaking of the fast. At 1900 on 24 October 2005 during the special Eid Al Fatr

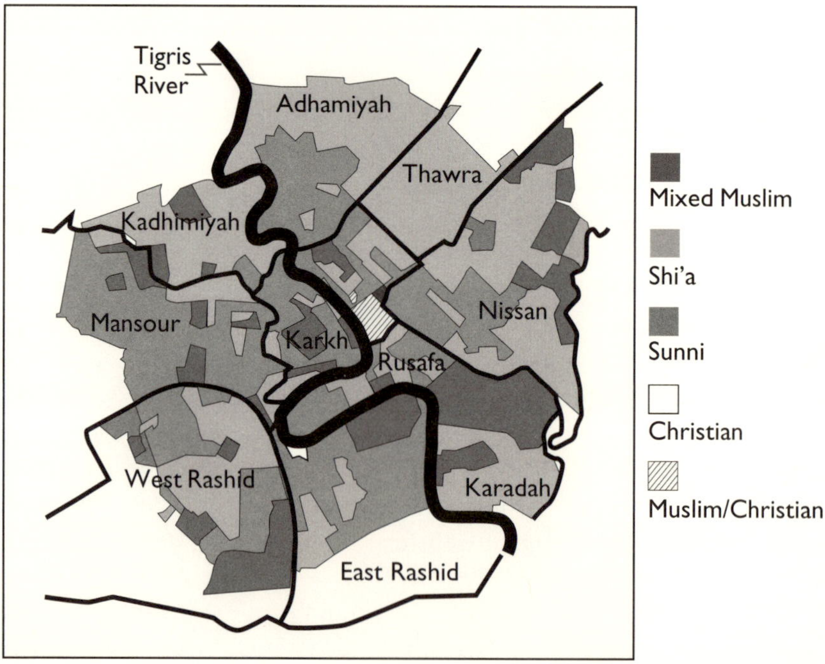

Figure 3. Baghdad sectarian map

prayer, Muqtada Al Sadr appeared on TV station Al Hurrah to give prayers followed by a speech. He stated that it should be considered an honor for any members of Jaysh Al Mahdi to be captured or killed by "occupation" forces—the ultimate form of struggle against the "occupiers."[13]

Traditionalism Versus Fundamentalism. In January 2006, when the Mujahideen Shura Council was formed, a statement on a jihadist Web site announced that Abu Musab Zarqawi had stepped aside as "emir" of the council in favor of an Iraqi named Abdullah Rashid al-Baghdadi.[14] The switch in leadership came from the council's leadership discontent with Zarqawi's targeting of civilians and the seemingly indiscriminant level of violence against the Muslim brothers. The council wanted killing of the Americans as the primary objective. Zarqawi did not; he was on the hunt, and everything was fair game. Foreign fighters, vanguard troops had been recruited for the war in Iraq and arrived for the blind destruction against the coalition and the citizens of Iraq. Allied in this offensive were the fundamentalists, the true believers who held a singular goal to subjugate the people to Islam.

Along with Zarqawi, the campaign to subjugate the population to Islam was on. This second change did not manifest itself until March, when a wholesale wave of attacks on the civilian populace began to take its toll. Visibly evident was the female citizens' change, throughout the city, from wearing western-style clothing to traditional abayah dress.

The Shi'a Counterattack. After being deposed from the council, Zarqawi commenced an independent action designed to retain his credibility as a grand emir by refocusing efforts against the Shi'as. The attacks were effective. So substantial were the assaults, the Shi'a organized their army—the Jasyh al Madhi— to defend by counterattacking Al Qaeda and Sunni forces. Zarqawi's success brought on not only the third sea change but also his downfall. Small pockets of battlegrounds began to emerge throughout Baghdad. Religious hatred exploded one night and stayed that way for the months to follow. Hayys (wards) in Mansour, Doura, New Baghdad, and elsewhere throughout the city went into play. Shi'a militias began clearing Sunni areas throughout Baghdad. Sunnis either stayed, fought, or fled. The resulting actions started the reconciliation of Sunni and

Shi'a boundaries in Baghdad to establish religious whole neighborhoods in the city. On 7 June, Zarqawi's terror came to an end as an U.S. Air Force F-16 dropped two 500-pound bombs on his position.

Through the Backdoor. In May and June, the fourth sea change emerged as Iranian subversion became quite apparent. There was no singular event that marked the Iranian emergence into Iraq, just a continuous drumbeat of Iranian influence. Constant reporting of Iranian agents infiltrating into offices of the newly elected government, visits by Government of Iraq officials to Tehran, and promises of Iranian funding of Baghdad city reconstruction. These are examples of several encroachments by Iran into Iraq. In actuality, what was occurring was a natural progression of influence based on a perceived vacuum of power during two periods of time: a) the flux when the new Government of Iraq was established and b) when Iraq could exert its own control of the nation. The Iranian government in the interim was exerting influence in order to obtain regional objectives and protect its western flank.

Tribal Warfare. In July, the fifth evolution came. Shi'a counterattacks became the Shi'a offensive. The task at hand was to clear Sunni neighborhoods across Baghdad, for the purpose of solidifying Shi'a control of not only the primary city in Iraq but the seat of national Iraqi power. A second effort went into action to clear the north to south highway between Baghdad and Najaf, Karbala, and Hillah. This clearing would ensure that the pilgrimage lines to the Shi'a holy places were open and protected and just as important, so were the lines of commerce. This fervor took time to build, but by August, was gale force. Death squads and death companies ravaged the city with assassinations rising as high as 100 each day. The purpose of the offensive was to *reconcile* tribal boundaries.[15] Mistaken by many, as sectarian violence, it really was a confederation of tribes supported externally by neighboring tribes against another amalgamation. This phase of the Battle of Baghdad had Sunni Janabi, Zubaydi, and Jiburi tribes pitted against a more homogeneous and advanced Shi'a tribal structure codified under the Sadr and Badr organizations. *Reconciliation* was advancing.

Following and sometimes in conjunction with reconciliation was *federalism*. The pinnacle moment came when the Iraqi Council of Representatives (CoR) voted for federalism.

> In the CoR meeting of today, 11 October 2006, a reading of the United Iraq Corporation (UIC) draft of the Federal Region Formation Law (FRFL) was conducted and followed by a vote of passage of the draft. The FRFL passed with an uncounted simple majority hand vote. This occurred despite the boycott of the session by the Sadr Bloc, Tawafuq, the National Dialogue Front, the Fadilah Party, and the Liberation and Reconciliation Front, all of which could muster only 105 seats of the 138 seats needed to prevent a quorum. It appears that deserters of the anti-FRFL front must include members of the `Allawi coalition (25 seats) and the Da`wah Party (15 seats). The pro-FRFL front managed to attract 33 deserters from the anti-FRFL front.[16]

According to the terms of the new Federal Region Formation Law, provinces that wish to form a federal region may commence proceedings from the date of passage of the FRFL in the CoR.

Commencement of proceedings refers to presenting the request for formation of the region to the Council of Ministers. This council in turn shall refer the request to the Independent Electoral Commission of Iraq (IECI) to arrange a referendum on the request. The FRFL states that the region shall form upon approval by a simple majority of members of the electorate of the proposed region who actually cast votes in the referendum. Parties who desire to engage in federalist undertakings are now free to do so. The parties who oppose the federalist agenda are likely to show extreme anger, quite possibly accompanied by violent repercussions, at this defeat.

The future of federalism will play out over the course of 2007, but it is now set in motion; and the formation of "Sunnistan," "Shiastan," and Kurdistan has settled on a firm foundation of Iraqi law.

As reconciliation and federalism progress, expect a final phase of consolidation where the transition of majority rule will go to the reigning Shi'a force. The decision to be made is whether Iraq will be ruled by a pro-Iranian Shi'a controlling force or an Iraqi Shi'a power in charge.

The five evolutions of the enemy demonstrated the violent cross currents that MND-B encontered in 2006. In many cases, each evolution was difficult to predict, and in many cases only understood after MND-B was well into the fight. In all cases, before launching any civic-action operation to counter the change in enemy activity, the MBD-B G9 staff reviewed the fundamentals. The theory and fundamentals behind civic action are covered in the next chapter.

4. Theory of Civic Action

> The specific focus of military civic action is to create a favorable environment in which to support current or anticipated operations and to gain the support, loyalty, and respect of the people for their local and national government by assisting communities in conducting health, welfare, and public works projects; improving the economic base of the nation and standing of the host country and allied forces with the population.
>
> — FM 31-73 Advisor Handbook for Stability Operations

Civic action is designed to reduce or remove the causes that fuel the insurgency. The subversion of a government by the insurgency is the battle fought to wrestle control of government away from the seated authority by demonstrating to the people that their leaders cannot protect them, provide essential services to the citizen at large, or offer economic opportunities for a better future. The insurgency proposes to offer the "oppressed" a "better" future and lures the people away from the government by undermining it. The objective of civic action is to synchronize military force with interagency and international partners to not only build trust and confidence by the people for their government and allies but also erode popular support for the insurgency. Civic action targets the COIN center of gravity.

The insurgency proposes to offer the "oppressed" a "better" future and lures the people away from the government by undermining it.

The center of gravity for a COIN is the populace. Specifically, three components exist:

a. Tactical center of gravity is the adult population (i.e., 26 years old and older)—the antigovernment, neutral, and progovernment population base.

b. Operational center of gravity is the next generation (16-25 years old)—the recruiting base for the next wave of insurgents.

c. Strategic center of gravity are the children (15 years and younger), those who will inherit a nation either at peace or war.

The insurgent fights for control of the people, and the government fights to protect the people. Separate the people from the insurgents and you win; if you don't, you will loose.

The civil assessment of enemy intent conducted by MND-B staff was that elements of the insurgency would act as follows:

a. Conduct underground activities to foster political discontent with the national, provincial, and local government in order to continue to subvert the progress of democracy and set conditions for an insurgent selected form of government based on Sharia Law.
b. Engage in financial activities to support the insurgency, allying with criminal actors.
c. Mostly through coercion, force local nationals to become an auxiliary to the insurgency to support the enemy operations and give sanctuary.
d. Conduct guerrilla operations against coalition and government security forces to degrade/destroy the people's trust in the government's ability to protect and provide essential services to the citizen.
e. Transit from staging bases outside of the cities along the major and secondary roads to caches sites, then onward to safe houses within the city, conduct attacks, link up with underground members for payment, and then return outside the city.
f. Usher in externally supported government agents and foreign fighters who desire to keep the country destabilized, or turn the host nation into a proxy or annex to the hostile nation.
g. Subjugation of the populace is the means to achieve the enemy's end state.

Civic action has three goals: a) gain the support of the populace in order to disrupt enemy movement and sanctuary, b) divulge the location and disposition of insurgent leaders and forces, and c) degrade or eliminate the insurgent logistics base. Civic action removes the ability of the insurgent to hide openly amongst the populace and denies or degrades the insurgent access to locally provided supplies, provisions, and services. Civic action also allows for coalition, allied, and host nation army and police forces to gain local information to insert in the precision-strike targeting process.

To conduct an effective civic-action program, the insurgent's advantages must be attacked. The insurgent operates from a "home field" advantage, and he knows it. The coalition operates for a defined time period and will leave, but the insurgent—whether home-grown or a foreign fighter—will always be there. Therefore it is difficult to separate the people from the insurgent, when the people know that one day they will face those who make up the insurgency when coalition forces are not present. The same can be said when coalition forces conduct only day-time operations and return to base camps in the evening, and the people must face the insurgent at night with no coalition security. Separating the populace from the insurgents when the people fear for both short-term and long-term security is a complicated operation that requires much skill.

The purpose of civic action is to turn the people against the insurgency in order to reveal the enemy's location, disposition, and intent. We want the locals to come out and divulge the location and disposition of guerrilla forces, the underground, and the auxiliary, so we can kill or capture them and protect the populace.

We want the locals to come out and divulge the location and disposition of guerrilla forces.

The method involves focusing civic action to build and maintain civil cooperation by executing short duration civic projects that have a high impact on the quality of life for the populace and provide outlets for the people to provide leads and information to civil authorities and security forces. Large-scale, long-term civic action will be used to build trust and confidence in local and provincial government and are best managed by a single project office outside the military, but managed on a streamlined process to ensure completion as promised.

The end state is to influence population centers that are pro-coalition to stay friendly and provide information leading to the kill or capture of guerrilla forces, the capture of underground leaders, and the degradation of auxiliary support to the insurgency. Civic action is conducted in neutral areas to influence them to become friendly and move hostile areas to becoming neutral.

Every civic action has a military objective and does not seek to execute random acts of kindness or supplant the responsibilities of the Department of State. Civic action manages the civil side of the COIN operation, tailored against the enemy's order of battle. This order consists of the armed insurgent (the guerrilla), the political and financial arm of the insurgency (the underground) and the civilian support to the insurgency (the auxiliary). The overarching construct is to separate the guerrilla from the populace (turn the populace against him and clear the guerrilla), occupy and reduce the underground (the democratic process), and employ the auxiliary (restoration operations and infrastructure development).

Civic action is centrally resourced with decentralized execution. It focuses on identifying the causes of instability, which promulgate the insurgency, and high impact, short-term projects and sustainable plans to counter it. Civic action develops protection plans for the local population (curfews, local security force support, neighborhood watch, and protected zones) and organizes strategies to turn hostile locals friendly (black to white), neutral populations to supportive (gray to white), and retain positive support. Additionally, civic action is the cornerstone to reduce lower level guerrillas and separate the foot soldier from the insurgent's leadership. Key tenets of civic action include the following:

a. Balance strike operations to kill or capture the enemy with civic action to build civil cooperation and trust and confidence with the populace.
b. Establish a security environment that facilitates the development of local economies, promotes good and fair governance, and the provisioning of basic public services.
c. Separate the guerrilla from the populace by turning the people against the insurgency.
d. Facilitate the transfer of battle space to indigenous security forces and civil self and independent rule.
e. Support the indigenous political process along basic democratic lines.
f. Support engagement of host-nation leaders, tribes and clerics.

g. Transition military government functions and civic action to the U.S. Department of State internal development programs.
h. Support local and district governments in sustaining the local level delivery of basic needs and essential services.

Execution of civic action is both operational and tactical. Operational civic action encompasses support of the Department of State and transitioning the economic and governance lines of operation to the Department of State, in advance of the security situation becoming passive. Typical operational and tactical actions are shown in Table 4.

Table 4. Execution of Civic Action

Level	Civic Action
Operational	Engage local, district leaders and populace in obtaining tactical objectives.
	Support development, maintenance, and exploitation of positive relations between the military and the civil community.
	Manage the civil populace and the impact of military operations on the civilian populace.
	Gain support from civilian agencies of the U.S. government, local, and district governments of the host nation, civilian agencies of other governments, and nongovernmental agencies in order to obtain tactical objectives.
	Exercise temporary military control of the civilian populace in specified areas until control can be returned to local and district government authority.
	Provide assistance to meet short-term, temporary, and life-sustaining needs of the civilian population.
	Provide expertise in civil-sector functions in order to assist the local and district government in developing and restoring civil-sector capabilities and functions.
	Train the host-nation military to plan, train for, and conduct military civic action in order to support development of the military and build the legitimacy of the government.
	Plan, execute, and manage projects to improve quality of life and improve government's ability to govern, specifically focused on the governmental functions of public administration, public health, public safety and public utilities.

Table 4. Execution of Civic Action (continued)

Level	Civic Action
Tactical	Secure auxiliary support to the insurgency in order to disrupt enemy lines of communication into the cities. Occupy the underground in order to disrupt fiscal and technological support to the insurgency. Create a rural auxiliary to divulge guerrilla activity in sector in order to disrupt guerrilla movement and sanctuary. Create a local population base willing to divulge insurgent activity in sector in order to disrupt enemy movement, resources, and sanctuary. Focus on ascertaining the location of the underground in order to disrupt subversion.

Related civic-action mission sets encompass critical ancillary operations such as humanitarian assistance, internally displaced person management, consequence management, and interagency operations.

Humanitarian Assistance (HA)

HA encompasses short-range programs aimed at alleviating present suffering. This assistance is usually conducted in response to natural and human-made disasters, including combat. The assistance provided by military forces is limited in scope and duration, and it must not duplicate other forms of assistance provided by the U.S. Government. It is designed to supplement and complement the efforts of the host nation civil authorities or agencies that have primary responsibility for HA. The assistance includes programs conducted to relieve or reduce the results of disasters or other endemic conditions such as human pain, disease, hunger, or privation that might present a serious threat to life or result in great damage or loss of property. HA operations can encompass both reactive programs, such as disaster relief; and proactive programs, such as humanitarian and civic assistance.

> Task: Conduct HA operations as short-range programs aimed at alleviating human suffering.
>
> Purpose: HA operations are designed to supplement or complement the efforts of the host nation civil authorities or agencies to offset humanitarian strife and avert depleting/degrading combat power.

HA programs are designed to assist civil authorities in carrying out their responsibilities of providing assistance and basic services to the local populace. Typical HA tasks are as follows:

a. Execute HA operations according to coordinated plans and guidance.
b. Employ sufficient combat power to safeguard the populace, defend the perimeter, escort convoys, screen the local populace, and assist in personnel recovery operations.
c. Assess HA needs, security, food, water, medical, and shelter.
d. Identify HA resources/agencies available to support operations and establish contact and working relationships.
e. Publicize HA plan through local public information operations and tactical psychological operations (PSYOP) assets.
f. Execute operational plans for immediate mitigation of political, economic, legal, social, and military issues associated with HA activities.

Internally Displaced Person (IDP)

An IDP is defined as any person who has left his residence by reason of real or imagined danger but has not left the territory of his own country. Any persons or groups of persons who have been forced or obliged to flee or leave their homes or places of habitual residence, in particular as a result of or to avoid the effects of armed conflict, situations of generalized violence, violations of human rights or natural or human-made disasters, and who have not crossed an internationally recognized state border (United Nations definition contained in the guiding principles on internal displacement).[17]

> Task: All units report status of IDPs and camps.
>
> Purpose: Battle track IDP locations and movement in order to avert any humanitarian catastrophe, determine indications of sectarian strife, and build a common operating picture of IDPs.

The staff must determine an appropriate response, with a focus on security, food, water, medical care and shelter, accurately monitoring friendly and enemy dispositions and notifying appropriate units and individuals. Critical IDP tasks are as follows:

a. Identify requirement to initiate IDP operations.
b. Plan and deconflict IDP routes, collection points, assembly areas, and camps and post on operational graphics.
c. Assess IDP needs, security, food, water, medical, and shelter.
d. Coordinate IDP control measures with adjacent units to minimize redundancy and increase efficiency of IDP operations.
e. Publicize IDP plan through local public information operations and tactical PSYOP assets.
f. Execute operational plans for immediate mitigation of political, economic, legal, social, and military issues associated with IDP activities.[18]

Consequence Management (CM)

CM involves actions taken to maintain or restore essential services and manage and mitigate problems resulting from disasters and catastrophes, including natural, human-made, or terrorist incidents.

> Task: Plan for and execute CM operations as an integrated part of all combat operations premission activities. Additionally, CM may involve prepositioning supplies, services, resources, and materials to rapidly mitigate any physical, psychological, cultural, or emotional damage to the civilian populace. Effective CM operations preempt or deter the discontent of the populace with coalition forces and support the people's confidence in their government's ability to protect them.
>
> Purpose: Prepare for and respond to attacks against the civil populace, conduct immediate CM operations in order to return the affected area to preattack conditions and mitigate the effects of the attack. Immediate is defined as coordinating repairs within 48 hours of the damage.

In general, CM operations can be as complex as deploying a task force to specially mitigate the effects of the attack or more direct such as providing monetary reimbursement. In general, actions for the CM objective are focused on conducting a damage survey, determining the immediate needs, documenting the damage, and providing an initial mitigation plan. The acronym INDIA (Immediate

Needs, Damage, and Infrastructure Assessment) describes the steps required for documenting damage and reporting requirements to the next higher headquarters in order to execute any of the following CM options:

a. Prepositioned assets mitigating damage at the point and time of impact. Repairs must fit the unit's ability to secure the location during the CM operation.
b. Claims cards and reimbursement instructions provided to the local populace in order to process for damages. Claims cards are ideal for CM when the situation does not permit lengthy loiter time on the objective or return.
c. Contract with a local vendor on an indefinite delivery/indefinite quantity contract to conduct repairs. The local contractor will photograph the damage, provide an estimate, and repair the damage as directed by the coalition. This method allows for a local entity to conduct follow-on repair without forcing a return of coalition forces.
d. If the situation warrants, the commander may constitute a task force tailored to mitigate the effects of the attack. For attacks against civilian targets by enemy forces, this generally involves coalition forces providing security, transportation, supplies, rubble/debris removal, and the ability to repair damage in support of the local government.
e. Commander's Emergency Response Program (CERP) funding can be used to support CM requirements. The type of support available consists of CM packages or local vendor repair. CM packages consist of prestocked supplies stored at a Forward Operating Base, such as construction supplies for restoration operations.

Interagency Operations

Interagency operations are the most critical part of a COIN campaign. A unilateral military operation means unilateral failure. All elements of national power—diplomatic, information, military, and economic—should be employed toward defeating the enemy.

Interagency operations can be defined as the coordination, synchronization, and decisions that unify the diplomatic, informational, military, and economic efforts into mutually supporting action. In other words, interagency operations deconflict U.S. Department of State governance programs, U.S. Agency for International Development (USAID) initiatives, and U.S. Information Services (USIS) activities with military offensive, defensive, and stability operations. Hence, interagency operations become the driving operational level consideration in COIN operations. The keys to interagency operations are communications, relationships, and power sharing.

The operations are not resident at the embassy level alone. The myriad of organizations that populate the battlefield range from non-governmental to coalition government groups as well as those within U.S. Government agencies, local, provincial, and national government. In order to ensure synchronized operational level COIN action by all elements of national power, the U.S. country team serves as the standing body to coordinate multiagency, intergovernmental, and international efforts.

The method for interface between the tactical unit and the local government organizations generally comes through face-to-face encounters to capture local needs/requirements—information used for needs-based project nominations. At the higher headquarters a civic-action group can synchronize project nominations to ensure unity of effort and economies of scale. To integrate tactical civic action and project nominations with the highest level of the host nation, typically a general officer chairs a joint planning commission that acts as follows:

a. Presents project nominations.

b. Receives updates on project issues and completions.

c. Serves to connect local to provincial government, as well as deconflict civic actions with the provincial council, Department of State, industry, and other participants.

Hence, civic action and its subcomponents are the other side of COIN. Its driving elements are the military objective, building the people's "trust and confidence" in their government, civil information that leads to the location and disposition of the enemy, civil cooperation that moves the populace to active or neutral support of coalition forces, and coordination of the interagency effort as paramount in transitioning the economic and governance activities to the Department of State.

Command and Control of Civic Action

When an insurgency forms, the affected populace is too dispersed to be centrally controlled. Local actions can take place to manage specific population centers, but the key to subverting an insurgency comes from widespread civic action.

For civic action, control is mostly bottom up, with only priorities, parameters, and resources flowing from the top-down. The higher command and control element, which is not engaged in active combat operations, is responsible for establishing priorities for the subordinate commands to synchronize the actions across a wide enough area to disrupt enemy operations and deny them freedom of movement. Parameters are established to delineate the subordinate command's level of authorities and define the limits of their operations, so that the focus of military operations becomes protecting the populace. Civic-action projects are distributed to weight the battle,

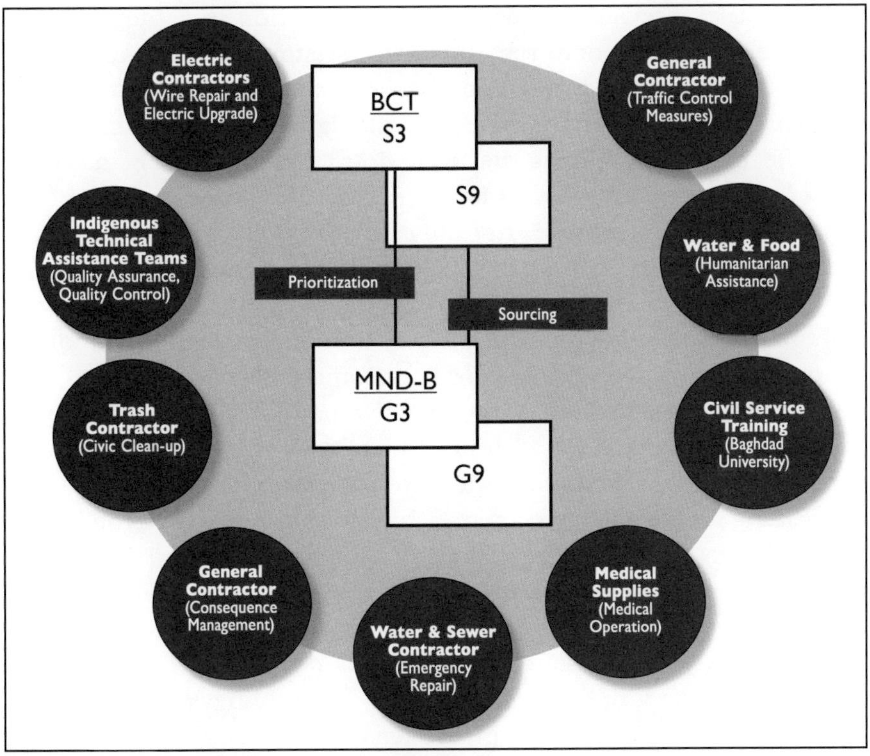

Figure 4. Contract resources coordination

systematically destroy, disrupt, or degrade the elements of the insurgency, (i.e., guerrilla, underground, and auxiliary).

The basic premise of civic-action operations is that the lower echelon of command is where the decisive operation occurs. The key leaders are platoon and squad leaders, and the fight is over the populace. In fact, the operational environment is far too complex and localized for an overarching mechanism to synchronize small-unit action. Deliberate operations and targeting coordinated at the divisional level and transmitted through the plans and orders process is an institutional impediment. Division staffs tend to divert staff energy to try to "mass nonlethal effects" to coordinate a ribbon-cutting ceremony at a refurbished school for an infantry company. A division staff's focus is 96 hours and out for planning and resource allocation, not 96 hours and in for resource management. What is needed is the higher headquarters front-loading manpower, funding, equipment, and enablers to conduct the operation, then sustain the effort.

The other prevailing reason why civic action is a tactical fight supported by operational resources and sustained by strategic purpose is the composition of the populace. Areas of economic stress and humanitarian need produce the guerrilla fighter. The underground is very effective in convincing the "have nots" that they can be a part of those who "have," but only by force of arms. The demographics of the combat zone demonstrate that only 2 to 3 percent of the populace are affluent, 20 to 35 percent are middle class, and the remaining 68 to 72 percent are lower class (the "have nots"). Baghdad's considerable middle class prior to the war has diminished—been killed, sought refuge outside the city, or is in some form of hiding.

With such a divergent and expansive populace, the optimal way to remove the home-field advantage from the insurgent is to work in their backyards and conduct wide-area warfare. An example is commissioning a series of civic actions, based on the needs of the locals, targeted into the area where guerrilla forces are bound to operate (the dead space), which is usually the lower income, impoverished areas. In general, the intensity of the insurgency is proportional to the expanse of the neglected populace. As demonstrated during the Battle of Baghdad, one dead space were several hayys (wards) in the Rashld beladiyah (borough) in the east side of Baghdad. A lesson learned for clear-and-build operations in areas such as Rashld was that one combat arms battalion can clear a 24- to 36-square-kilometer urban area consisting of two- to three-floor residences, and one Brigade Special Troops Battalion is required for the build phase.

The overarching command agent for civic action is the commanding general or the commanding officer. The commander of the higher headquarters interfaces with his subordinate commanders and articulates the prioritization, parameters, and resources available for his field commanders to execute the civic-action program. The commanding general or commanding officer in many cases will appoint his deputy as the primary command group official responsible for direct oversight of civic action. The staff proponent or principal staff agent responsible for the day-to-day management of civic action at the division level is the assistant chief of staff, G9, while at the Brigade Combat Team and battalion level is the S9.

The greatest control mechanism available for civic action is money. Cash rents loyalty, and paid friendship leads to friendly action; in fact, the more money spent in a single area, the greater the friendship. Such concepts, discussed thus far, guided our civic-action planning and operations throughout 2006.

5. From Theory to Practice—the Civic-Action Program

Closing in on Baghdad involved putting research and theory into practice. The Red Zone analysis—the intelligence preparation of the operational environment—had served its purpose:

a. Determined the causes of instability, anticipated prior to deploying for Operation Iraqi Freedom 05-07
b. Derived why the enemy was focused on Baghdad.
c. Divulged what made the populace side with the insurgency.
d. Determined what the religious, historic, cultural, and economic undertows were beneath the surface.

Arriving in Baghdad further refined the assessment due to finding a codification of the causes for instability. A lesson learned was that the behavior of American soldiers toward the citizens of Iraq was abrasive. The overflow of Forward Operating Bases and the force protection measures in and around Baghdad had made the city not only walled but under siege. No connectivity between the District and Nahia Advisory Councils to the Baghdad Provincial Council existed, hence no continuity of government or people's representation. The infrastructure lacked in many areas and was failing throughout. The U.S.-sponsored reconstruction projects were causing huge discontent with the populace because all the promises made for electricity, fresh drinking water, functioning sewer systems, and a new landfill were not kept—far from completion or would never be completed.[19] The institutions of the U.S. Government had settled in, and progress was severely hampered as American bureaucrats used peacetime procedures and practices to manage a COIN campaign. Another discovery was a great inability to keep up with the flow of information being received by a multitude of sources and an ever painful compulsion to place local action in American terms, vice understanding how the indigenous populace thought and accepting why they did what they did for their reasons, not ours.

. . . progress was severely hampered as American bureaucrats used peacetime procedures and practices to manage a COIN campaign.

Entering the Battle of Baghdad also meant reevaluating the plan. The developed action scheme had a specific task and purpose for civic action. The task was to develop "trust and confidence" of the people with their government or to improve the quality of life to lift the populace up to Maslow's hierarchy of needs. The core reasons behind the civic-action program were to build civil information, leading to the location and disposition of the enemy, and civil cooperation in support of coalition forces achieving the command's objectives. The purpose of the program was to deny the enemy the populace, which we considered to be the center of gravity.

Major Doug Davids took the lead in validating assumptions and ensuring the plan of action could achieve the task and purpose in the operational environment. He set forth on an area-wide survey to determine the core tenets of the civic action. He went to Babil, throughout Baghdad, and worked interagency integration. Davids derived the themes of the civic-action plan following 4 out of the 16 governmental functions.[20] From his survey the civic-action plan transitioned to a civic-action program.

MND-B planned and executed a civic-action program for a 62,000 combined force consisting of 27 U.S. and Iraqi brigades engaged in active COIN operations. The program formed the day-to-day underpinning for a population consisting of 11 million civilians across a 16,664-square-mile area and served as the primary force to stabilize the MND-B area of responsibility. The program also consisted of 1,044 completed or ongoing projects at a total value of $138 million, executed by seven U.S. Brigade Combat Teams. Components of the MND-B civic-action program were included in the following activities, most of which are further detailed herein:

a. Operation Glory Light, the main effort for the division's COIN operation in January, centered in south Baghdad
b. Operation Willing Eagle, the permissive entry by the Ministry of Electricity to validate that the off-limits Yusufiyah Power Plant was a key enemy base of operations
c. Operation Baghdad Is Beautiful, the city-wide cleanup program
d. Operation Harvest Light, the date-palm spray campaign in support of the Ministry of Agriculture

e. Operation Together Forward II, the infrastructure rehabilitation in focus area throughout Baghdad
f. One Civil Affairs company to the Baghdad provincial reconstruction team (PRT) and another to the Babil PRT with lieutenant colonel (O5) liaison officers
g. Command oversight on civic action, ensuring that the commanding general, deputy commanding general, support, and chief of staff had full visibility of the expenditures, face-to-face engagements, and societal impacts of the division's civic-action program
h. Reporting and command oversight of the CERP, revamping by implementing the weekly civic-action group—the Brigade Combat Teams providing direct accountability to the division's deputy commanding general, support, and chief of staff via the Command Post of the Future
i. Reconstruction reporting, converting to the Iraqi Reconstruction Management System with 3,423 total records for fiscal years 2005 and 2006.

Figure 5. R3 Water Treatment Plant, Muhallah 329, Baghdad, 22 July, COL Tom Vail and LTC Chris Hall, 506th Regimental Combat Team

Measures of success for MND-B's civic-action program in the economic development effort included the date-palm spray campaign, a

$750 million economic impact that contributed to a 21-month low in unemployment, and the opening of 65-80 percent of local business in the focus areas. Success occurred in improving governance with interagency synchronization, building local government agencies, and then connecting local-to-provincial government leaders. The cost was $12.55 per person.

What follows is a synopsis of civic-action programs—a description of a wide range of initiatives taken to engender civil cooperation in Baghdad in 2006. Through these actions, information about insurgent forces was obtained, infrastructure was refurbished, and increasing civil authority was transitioned to local governments.

Operation Glory Light

Soldiers from the 6th and 9th Iraqi Army (IA) Divisions, supported by MND-B soldiers, conducted operations to gain control of a terrorist stronghold south of Baghdad. The operation began in the early morning hours of 2 March, when the soldiers of 3rd Battalion, 1st Brigade, 6th IA Division and 1st Battalion, 502nd Infantry Regiment, 2/101 AASLT air assaulted into the Sadr-Yusufiyah area. Following the air assault, the soldiers were attacked by indirect and small-arms fire. They also discovered a roadside bomb.

Soldiers from the 6th and 9th IA Divisions then moved into position to clear, search and secure roads and houses in the town. From 5-6 March, the 9th IA Division uncovered nine weapons caches, consisting of 131 mortar rounds, nine 50-mm mortars, 17 rockets, two rocket-propelled grenade launchers, six AK-47 rifles, two BKC machine guns, and a large amount of bomb-making materials. The operation lasted for 1 week.

Figure 6. An IA soldier stands guard during the search of a house as part of the combined operation with coalition forces; U.S. Navy photo by photographer Mate 2nd Class Samuel Peterson

The purpose of Operation Glory Light was to deny terrorists sanctuary and preempt enemy attacks in the Baghdad area. The key tasks were as follows:

a. Conduct focused on intelligence, surveillance, and reconnaissance to accurately determine enemy disposition prior to execution.
b. Control routes that anti-Iraqi forces can use to withdraw from or reinforce.
c. Mass sufficient lethal combat power with the main effort to allow for a decisive overmatch of any enemy force.
d. Use fires to reduce enemy resistance before committing ground forces.
e. Secure LOCs into the objective area to maintain resupply and casualty evacuation operations by ground as a viable option.
f. Provide the enemy with multiple simultaneous problems; do not allow him to focus on a single axis of operations.
g. Be prepared to continue offensive operations south of Sadr Al Yusifiyah.
h. Conduct combined operations with Iraqi forces and maintain an Iraqi face on publicized operations.
i. Simultaneously conduct civic action and CM to mitigate the alienation of local nationals not involved in enemy activity.

Operation Glory Light established a Civil Military Operations Task Force that reopened two schools, treated 200 locals in medical operations, distributed 3,000 pounds of HA food, executed 10 localized projects at $1.7 million, discovered four caches, and moved a hostile qada (county of a governorate) to a neutral stance. Units conducted leader engagement and civic-action operations simultaneously with offensive operations to provide immediate CM and enhanced support for Iraqi Government and security forces, to include medical and veterinarian operations.

Civic-action teams conducted mitigation throughout the operation as well as assessment of public utility needs. Work was initiated on CERP projects in the Sadr Al Yusufiyah area. The civic actions included these activities:

a. Engaged in Shakariya, leading to reopening of two schools (16 Feb).

b. Combined interagency, Iraqi Police, and civic-action mitigation in Mahmudiyah of anti-Iraqi forces rocket attack (2 Feb).
c. Assessed emergency and essential services in Yusufiyah (28 Jan to 2 Feb).
d. Developed scopes of work for courthouse and two schools in Yusufiyah (1-2 Feb).
e. Developed scope of work for the Civil Military Operations Center in Mahmudiyah and traffic control points along routes.
f. Conducted key leader and populace engagement in Zaidon area to determine attitudes.
g. Coordinated Iraqi engineer assistance on civil military operations.
h. Assessed Mahmudiya City Hall and other possible projects for upgrade (24 Jan).
i. Distributed 3,000 pounds of food, 300 pounds of children's clothing, and 1,000 copies of the Baghdad *Now* newspaper to residents.
j. Treated over 200 patients for minor injuries and illnesses via 2/101 AASLT.
k. Assessed two schools, health clinic, water treatment facility, and a power substation for possible CERP project upgrades.
l. Coordinated water pipe repair through Nahia water minister.

The combined operation with 1st Brigade, 9th IA Division; the 4th Brigade, 6th IA Division; and the 2/101 AASLT led to the discovery of 26 roadside bombs, two vehicle-borne IEDs, and 15 weapons caches. The operation also involved seven engagements with the enemy, resulting in the detention of 16 suspected terrorists, two terrorists killed, and one wounded. Foreign fighters and insurgent leaders in Sadr Yusifiyah were killed or captured, the enemy organization in the Euphrates River Valley was disrupted, enemy supplies and equipment were captured or destroyed, and U.S. forces exerted control over the avenues of approach into and out of Sadr Yusufiyah.

Operation Tigris Waves

In March, elements of MND-B conducted Operation Tigris Waves to transform the town of Tarmiya, located approximately 15 kilometers north of Baghdad, from an enemy sanctuary to a place that self rules and does not actively support enemy activity. The operation was a textbook balance of *strike* operations and civic action that isolated the populace from the insurgency, created an internal capacity to govern and protect the citizens, and transitioned authority to civil control.

The plan for the operation had five phases that focused on isolating, occupying, and securing Tarmiya; providing civic action; and transferring to civil control. 1/4 ID was to secure the objective—prevent anti-Iraqi forces activity in Tarmiya from influencing Baghdad and eventually transfer security of Tarmiya to the local authorities, allowing coalition forces to maintain tactical over watch.

Key tasks were as follows:

a. Conduct continuous engagement with both civic leaders and the people to gain their trust and confidence.
b. Isolate Tarmiya—no one moves in or out of Tarmiya without our consent.
c. Occupy patrol bases inside Tarmiya with coalition and Iraqi Army forces to create a 24/7 presence.
d. Inject quick turn, high impact essential services projects that employ local contractors to help the average citizen of Tarmiya.
e. Create a wedge between the citizens of Tarmiya and the enemy through a combination of positive and negative reinforcement means.
f. Neutralize enemy once separated from the citizens of Tarmiya.
g. Establish an effective, fully manned Iraqi Police force that can secure Tarmiya.

The desired result included enemy forces cleared from the environs of Tarmiya, Iraqi Police assumption of security for the people of Tarmiya; Iraqi Army and coalition forces vacated patrol bases inside Tarmiya; and an end to the isolation of Tarmiya.

The civic action for Operation Tigris Waves was *Support to the Civil Administration*, specifically the establishment of civil control in the city. On 25 March, coalition forces met with local leadership at the Tarmiya Qada meeting to set the conditions for Iraqi Army and U.S. withdrawal from Tarmiya. The main focus was to establish a strong functioning police force that can secure Tarmiya for the public.

With Operation Tigris Waves in high gear, the 1st Brigade, working with Iraqis, conducted a medical operation (MedOp) in Tarmiya to remove or reduce local support for the insurgency. The goal was to improve the standard of living—offering a better future affecting strategic, operational, and tactical centers of gravity. With 375 local nationals receiving treatment, the Tarmiya clinic asked coalition forces to plan other MedOps in the future.

Operation Harvest Light

Operation Harvest Light, the date-palm spray campaign in support of the Ministry of Agriculture, was the first time the celebrated Iraqi date-palm orchards had been sprayed in 3 years. Date production had been a significant economic and cultural event for Iraq. At one time, date palms were their second largest export, generating millions of dollars into the Iraq economy. Producing approximately 30 percent of the world production in the 1960s, Iraq's dates were considered top quality. Consequently, the crop was a source of national pride for Iraq, and a productive crop promoted economic stability, generated revenue, provided jobs, and brought families back together.

Years of war and neglect reduced the number of palms from 30 to 15 million. Iraq has steadily lost its world market share to other countries. Because an infestation of the Dubas bugs will severely damage the crop if left untreated, the Ministry of Agriculture wanted to conduct aerial pesticide chemical spraying of about 75,000 hectares of date-palm trees in Baghdad, Babil, Diyala, Najaf, Wasit,

Figure 7. AN-2, Colt contracted to conduct aerial spraying—focus on the farmer and deny the enemy the dead space

and Karbala governorates, plus 75,000 hectares of wheat fields in Ninawa governorate beginning in mid-April.

The mission for MND-B was to support the Government of Iraq, Ministry of Agriculture aerial spraying of date-palm crops from about 20 March to 18 May in areas within Baghdad, Babil, Najaf, and Karbala provinces. The date-palm spray campaign was a Multi-National Corps, Iraq (MNC-I) wide civic-action project supported by the MND-B. It was the largest single economic impact and influence operation that the division conducted in 2006. The key tasks for MND-B were as follows:

a. Coordinate with MNC-I for the contract resources to spray the MND-B area of operations.
b. Organize the MND-B staff to support the operation and synchronize public affairs, PSYOP, logistics, and security actions.
c. Clear air space and inform ground units of the spray schedule.
d. Organize search and rescue, as well as recovery assets in case a contract aircraft was downed.
e. Obtain Government of Iraq approval to support the Ministry of Agriculture.
f. Coordinate political support in four provinces for the date-palm spray campaign.
g. Coordinate with the agriculture offices in four provinces.
h. Coordinate spray concentrations, locations, and chemical types across 208 square miles.
i. Conduct lateral coordination with adjoining divisions to transfer assets to their control in order to continue with the national date-palm and wheat crop spraying.
j. Monitor, report, and adjust spray operations.

The conduct of the operation was difficult. The date-palm spray campaign season is from 1 to 30 May in the Baghdad, Babil, Najaf, and Karbala provinces. MND-B and the Ministry of Agriculture (director generals of agriculture) synchronized spray zones, pesticide usage, and public information. At the same time, MND-B provided discreet project management in support of the Government of Iraq appearing to run the program. The target area (Baghdad, Babil,

Karbala, and Najaf) was about 43,000 (of 60,000) hectares of date palms and 100,000 hectares of wheat found throughout the Iraqi theatre of operations.

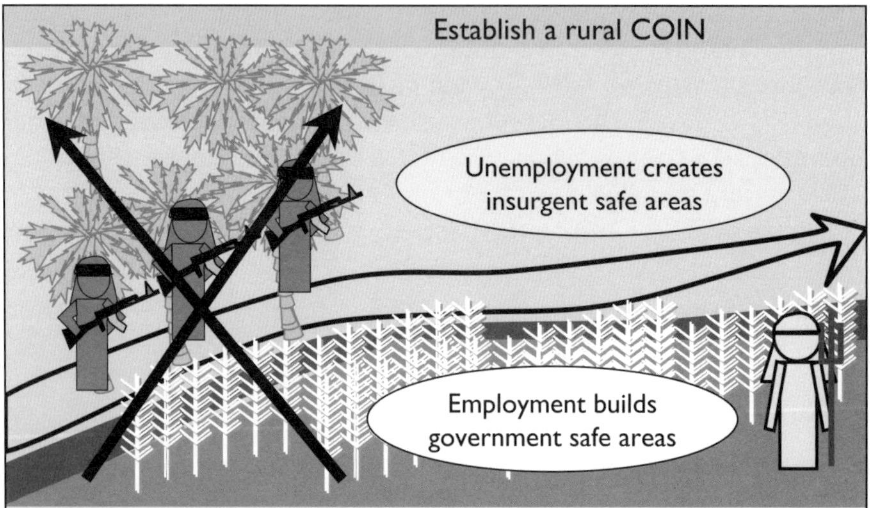

Rural COIN Tenet
Rural areas provide safe travel routes and havens for the enemy. Farmers are coerced to support the insurgent. Deliver agriculture action programs to employ the farmer, which moves rural support away from the insurgent and interrupts, disrupts, and denies key terrain to the enemy.

Figure 8. The rural COIN strategy

The contractor used Moldavian MI-2 helicopters and AN-2 Colt aircraft operated by English-speaking Turkish and Moldovan pilots to conduct the spraying. The contractor provided three fuel trucks. On about 20 March the contractor moved the six partially disassembled helicopters via truck, along with their three fuel trucks, aircrews, and logistics personnel from Turkey to the civil airport in Irbil. After the helicopters were reassembled and the proper Multinational Force-Iraq (MNF-I) approved avionics were installed, the contractor moved the aircraft in pairs to different airfields for conducting spray operations. Some of these airfields served as overnight sites, and some were to be used as logistics resupply locations. To track the progress of the spraying, transponders were installed on the

Figure 9. Aerial view of rural area

aircraft for recording their flight patterns. Here were the results of the effort:

a. Sprayed 54,450 hectares (134,548 acres or 210 square miles), which is 3.4 times larger than the District of Columbia (61.4 square miles)—specifically about 8.6 million palm and 8.6 million citrus trees (17.2 million total)
b. Killed an estimated 4,083,750,000 Dubas bugs, which Ministry of Agriculture sources said was a 75 percent infestation
c. Anticipated 651,412 tons of dates for the year's crop, 261,165 tons of citrus
d. At $120 per ton for dates and $110 per ton for citrus, estimated $110.9 million wholesale crop
e. Estimated export value up to $121.4 million wholesale based on the world average export price for dates, which translated into a potential of $562 million retail crop at five times wholesale
f. Based on two farmers per hectare, estimated that the operation employed approximately 108,500 from July to October to work the date and citrus crop.

The Ministry of Agriculture personnel were pleased with the effort. In Najaf date-palm spraying was well received, and full credit was given to the Ministry of Agriculture. Farmers had been warned to place a yellow flag high up in trees for areas that should not be sprayed to protect things like bee hives and fisheries, a worthwhile precaution.

In Taji, both Sunni and Shi'a were asked what they thought about date-palm spraying, and the answer was that they too were pleased. The people knew it would benefit them, suggesting that Iraq's agriculture should be one of the most important priorities of the new government. As for Babil, the locals were satisfied that spraying was finally happening again, yet unsure who was responsible for making it happen—the combined forces or the Government of Iraq. In contrast, the populace in Karbala was aware of the Government of Karbala's involvement in putting up obstacles to the spraying. That government's animosity and noncooperative attitude was expressed at that time towards coalition forces. Nevertheless, the spraying was conducted in support of the local farmers.

During the operation, MND-B had received indicators that spray zones along the Euphrates River in the vicinity of Taji were unsafe for aircraft due to enemy activity. The G9 informed the Ministry of Agriculture office that until the security situation permitted, the spray aircraft would not fly the Taji spray boxes. As a result, the Ministry notified the tribes and local governments in the area not to attack the spray aircraft or military aircraft accompanying them, so the areas could be treated. Because the temperature was

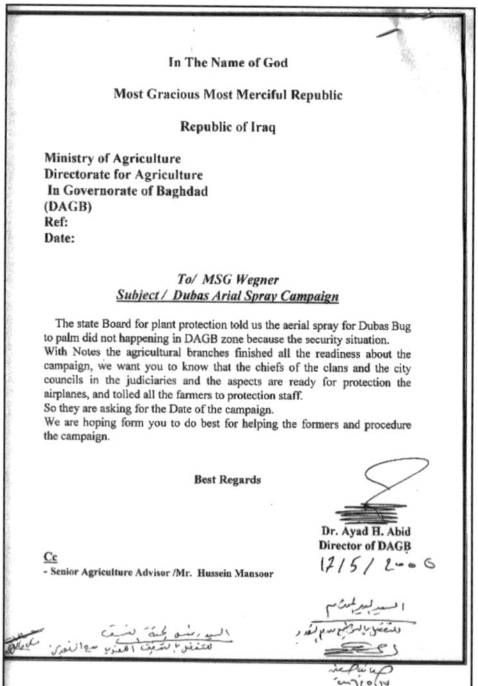

In The Name of God

Most Gracious Most Merciful Republic

Republic of Iraq

Ministry of Agriculture
Directorate for Agriculture
In Governorate of Baghdad
(DAGB)
Ref:
Date:

To/ MSG Wegner
Subject/ Dubas Arial Spray Campaign

The state Board for plant protection told us the aerial spray for Dubas Bug to palm did not happening in DAGB zone because the security situation.
With Notes the agricultural branches finished all the readiness about the campaign, we want you to know that the chiefs of the clans and the city councils in the judiciaries and the aspects are ready for protection the airplanes, and tolled all the farmers to protection staff.
So they are asking for the Date of the campaign.
We are hoping form you to do best for helping the formers and procedure the campaign.

Best Regards

Dr. Ayad H. Abid
Director of DAGB
17/5/2006

Cc
- Senior Agriculture Advisor /Mr. Hussein Mansoor

Figure 10. Letter from Ministry of Agriculture to MND-B

increasing, it was only a matter of days before the Dubas bug larva would mature and the eggs would hatch. A letter was written on 12 May from the Ministry of Agriculture to MND-B confirming that the aircraft would be safe.

The end result was that Operation Harvest Light built governance through economic development. In addition to widespread popular support with the farmers, trust and confidence were instilled in the provincial government's ability to serve its citizens. From Taji to Najaf, the local, provincial, and national governments were working for the people. The Ministry of Agriculture's office calculated pesticide concentrations and designated the spray areas. Officials from the Karbala and Tarmiya director generals of agriculture flew with the pilots to show them where to spray. District leaders informed the farmers of the spraying and governors and mayors were directly involved.

In summary, Operation Harvest Light served two purposes:

a. Being an influence operation, it moved guerrilla fighters out from underneath the date-palm groves where they cached weapons and explosives and from which they would stage forces.

b. Being an economic development program, the part-time guerrilla fighter was focused away from jihad against coalition forces to the date-palm trees.

Operation Baghdad Is Beautiful

Beginning in June, MND-B commenced Operation Baghdad Is Beautiful, a $5.2 million city-wide program for civic cleanup, consisting of 43 projects that reduced IED emplacement is in some areas up to six-fold and disrupted the IED network. The overall purpose of the campaign was to elevate citizen trust and confidence in their provincial government. The idea was to mitigate the constant reminders of

Figure 11. Civic-action cleanup program

war by removing rubble, excess wire, and barriers to demonstrate that the city was returning to prewar conditions. This migration was achieved in three ways:

a. Removing trash piles that could foster the spread of disease, therefore removing unsightly trash.
b. Reducing vector populations and demonstrating the benefits of clean neighborhoods to the local populace.
c. Reducing available concealment for emplacement of IEDs, thereby protecting the populace from attacks.

The project involved 15 areas in Al Karadah district, 3 areas in Babel district, 12 areas in Al Riyad district, 23 areas in Muhallah 903 district, and 7 areas in Al Jameea district. The desired result was to achieve a clean highway system in Baghdad with road markings and signs posted according to international standards. Designated city streets were to be clean and curbs painted, streets marked, and traffic and street signs emplaced. In addition, designated parks, medians, and the entrance to Baghdad International Airport were to be landscaped.

The local government designated 60 sites for cleanup, and work started on 29 May and concluded on 28 June. The actual work period was 21 days, and each day was 10 hours (total 73,500 hours). The contractors, local Iraqi companies, supplied transportation to and from sites, water, and soda at no cost to the workers. The average number of workers employed was 350 a day, and their average pay was $10 a day with supervisors getting $30 to $50, infusing approximately $75,180 in the short term.

The results achieved from the Karadah Peninsula cleanup effort included providing much needed jobs and contracts to Iraqi citizens, plus making a considerable dent in the detritus of war. The contractor hauled an average of 60 loads a day, utilizing 20 trucks and numerous loaders. The total number of loads hauled was about 1,200. Based on the number of loads, 11,345 tons of waste, trash, debris, barriers, and wire were removed.

Operation First Light

Operation First Light was directed by the commanding general of MNC-I in response to the default of a U.S. reconstruction project to build 142 primary health-care clinics. The background on the failed

U.S. reconstruction project is best described in an extract from the *Washington Post* article.[21]

> BAGHDAD – A reconstruction contract for the building of 142 primary health centers across Iraq is running out of money, after 2 years and roughly $200 million, with no more than 20 clinics now expected to be completed, the U.S. Army Corps of Engineers says.
>
> The contract, awarded to U.S. construction giant, Parsons Inc., in the flush, early days of reconstruction in Iraq, was expected to lay the foundation of a modern health-care system for the country, putting quality medical care within reach of all Iraqis.
>
> Parsons, according to the Corps, will walk away from more than 120 clinics that on average are two thirds finished. Auditors say the project serves as a warning for other U.S. reconstruction efforts due to be completed this year.
>
> Brig Gen William McCoy, the Army Corps commander overseeing reconstruction in Iraq, said he still hoped to complete all 142 clinics as promised and was seeking emergency funds from the U.S. military and foreign donors. "I'm fairly confident," McCoy said.

Figure 12. Al Thalith primary health-care clinic

> Coming with little public warning, the 86 percent shortfall of completions dismayed the World Health Organization's representative for Iraq. 'That's not good; that's shocking,' Naeema Al Gasseer said by telephone from Cairo. 'We're not sending the right message here. That's affecting people's expectations and people's trust, I must say.'

Six of the 49 clinics in the MND-B sector were between 90-99 percent complete. It was the Corps commander's directive to monitor and assist with security, and if required, construction and sustainment support to open these six clinics, as the U.S. Army Corps of Engineers completed the primary health-care clinics.

Operation First Light commenced on 21 May with Brigade Combat Teams conducting discreet site surveys, followed by a permissive survey to validate the level of completion. The teams would liaison with the clinic director and determine when the construction would be complete; whether any work was left undone; whether any medical staff and supplies were on hand; status of the services, water, electricity, and sewage to the site; and whether any obstacles remained in opening.

Coordination by the division with the U.S. Army Corps of Engineers ensured that the medical supplies and equipment were purchased and installed and the construction would be completed with Ministry of Health acceptance soon thereafter.

The strategy behind opening the primary health-care clinics was a forcing function of government. The projects were resurrected (after the Parsons pullout) and with the six clinics opening, the Corps commander wanted to ensure that the Ministry of Health was able to open, staff, and sustain the facilities. Over time, the six facilities opened and began seeing between 200 to 300 patients daily. This was the opening surge to areas that had not had local medical care. After a week or so, the clinic directors and the Ministry of Health asked coalition forces not to show up at the facilities because such visits would present a security or retribution risk to the clinic.

The Al Hurriya primary health-care clinic was the last one to open from the original six and that was on 27 September. The Ministry of Health has committed to providing medical staff and security personnel to these clinics as well as essential services that included water, fuel, and Class VIII medical supplies. Though other clinics remained to be opened, Operation First Light demonstrated that medical care could be provided at the local level, even when the Government was reluctant to do so.

Operation Freedom Light

In order to recognize the formation of the new Government of Iraq, MND-B conducted a large-scale release of approximately 3,000 detainees during June. This operation became a release program that would reintegrate detainees back into the civil society over the next several months.

The first release marked a milestone toward democratic rule and building a viable part of the Iraqi legal system in a new Iraq. The first Operation Freedom Light was staggered over a 5-day period, with the Prime Minister or select political/religious leaders participating in all of the releases.

To determine which detainees to release, a board reviewed 6,500 records that produced approximately 3,000 low risk individuals for release. Information about all detainees was entered into a database in order to identify repeat offenders if detained again. In June the recidivism rate was 5.5 percent according to unit data.

The concept of operations had buses escorted from the detainment facility to a release point in Baghdad. A central transportation hub, the Bus Station, was the avenue for released detainees to return to their homes. A reception team, to include Vice President Hashemi or local leadership officials, spoke with the detainees at the station to initiate the reintegration into the society. All detainees released were required to take an oath of allegiance to Iraq, the constitution, and foreswear all violence, after which the individual was released, generally to family members attending the release ceremony. The release was acknowledged by the Government of Iraq as an important part of due process.

Figure 13. Bus Station, public transportation for released detainees

This civic action was a classic reintegration program. It was succinct and efficient. The difficult part was the security operation, but the impact was that the Government of Iraq was able to demonstrate that the legal system was working and a far stretch from how Saddam would release detainees.

Operation Black Light

Operation Black Light commenced after a wave of terrorism hit Baghdad in May. The mission was to validate claims from open sources that religious-based killings were occurring. The methodology was to track morgue data, relate the findings to ongoing operations, and determine their effectiveness. With this information, an in-depth analysis surfaced ways to mitigate the effects of terrorist activity. Moreover, with the official data from the morgue, incorrect information about killings frequently found in the media was exposed.

Statistics published by the Baghdad Provincial Council indicate that terrorists were mainly targeting civilians, a likely cause for unrest and decreased stability. The primary terrorist methods were assassination (72 percent), IEDs (15 percent), mortars/rockets (6 percent), and car bombs (2 percent). With the exception of assassinations, markets and main streets were where most of the violence occurred.

Shootings comprised 94 percent of assassinations. Most victims were blindfolded and/or hands tied, which suggests execution-style murders (49 percent). Of the assassinations, 46 percent were targeted against civilians (rather than military, political, or religious leaders), which suggests the purpose was to intimidate.

Figure 14. Assassination numbers for May 2006 by type (1,105 total)

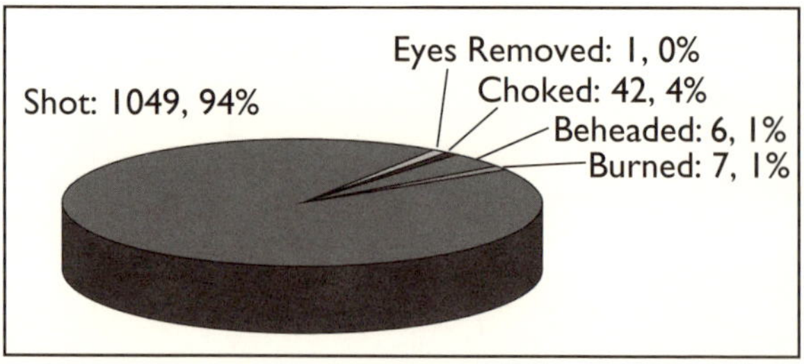

Car bombs were targeting civilians in popular gathering places, such as the market and the main street. These attacks were primarily used to cause instability in a specific area; the focus was to turn local residents against the police by placing doubt in their minds regarding police capabilities. Additionally, police were a primary target for terrorist shootings. In using car bombs, the terrorists' main tactic was using abandoned vehicles (vs. those of suicide bombers); because many such vehicles were in Baghdad, leaving another one on the street was not likely to draw attention. Moreover, it is more difficult to recruit people willing to die for the cause.

The average time of day for the terrorist activities was between 0900 and 1200, perhaps because of local residents gathering and shopping before the heat of the day. As for IEDs and shootings, most patrols started around 0900, and they can only vary their routes a finite amount of times.

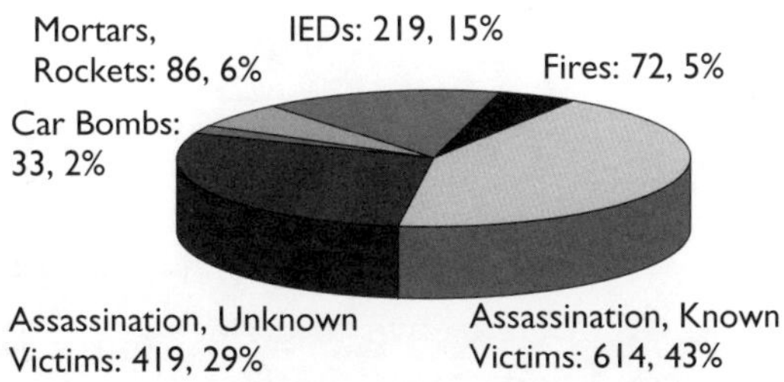

Figure 15. Terrorist acts for May 2006 by type (1,443 total)

The MND-B G9 coordinated a secure process of receiving monthly reports from the Baghdad central morgue statistics in order to conduct analysis and data verification. The data collected proved valuable in establishing a benchmark for evaluating overall operational effectiveness of MND-B and tracking mortality trends due to violence and disease. In addition, the information was used to verify related data, which was made public, and to document site-visit reports.

The ongoing analysis for Operation Black Light revealed that the security situation within MND-B was complex and fluid. Through September, deaths due to violence reached an all-time high, the nine-month total being 12,454. The morgue data suggested that four events contributed to the high death toll:

a. February bombing of the Samarra Mosque was the event that precipitated the 6-month increase in sectarian violence.

b. New Iraqi Government emerged in May as a Shi'a-dominated organization, leaving the Sunni's nothing but token power and the effect of increasing sectarian activity. The Shi'a-dominated Government asserting itself caused a struggle for power.

Figure 16. IEDs for May 2006 by targets (219 total)

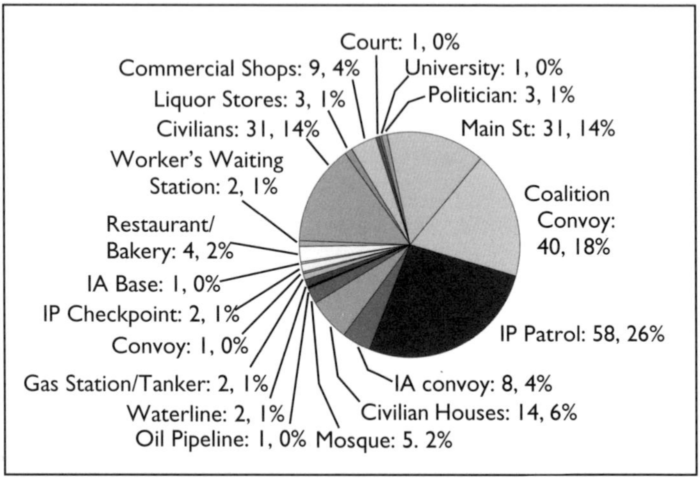

c. Once the coalition removed Abu Musab Al Zarqawi in June, Al Qaeda had reason to demonstrate their ability to operate without Zarqawi and began conducting increased sectarian operations, compounding the violence.

d. Operation Together Forward II in August was designed to minimize sectarian activity in the focus areas of operation but resulted in an unintended consequence of pushing sectarian operators into the outskirts and nonfocus areas around Baghdad, thus increasing sectarian activity in these otherwise quiet regions.

As the death rate dramatically increased, the obsolete equipment, understaffing, and poor operating procedures of the morgue overwhelmed the facility. The central morgue was designed to hold 36 sets of remains, six in each of the six refrigerated storage units. However, reporting indicated that not all of the storage units functioned, resulting in the original capacity becoming diminished. The morgue personnel reported receiving 8,000 bodies over the previous 6 months, an average of 44 bodies a day. The monthly morgue reports obtained by MND-B related similar numbers. Most equipment in the morgue was about 30 years old with the exception of newer computers and monitors. While reportedly 12 pathologists were on staff, many may have been part time. The facility was poorly managed, and the workflow was not orderly. Observers speculated that the morgue had not been cleaned in years, and no efforts had been made to maintain the facility.

Poor management, pressure from sectarian factions to not accurately report the level of violence, inadequate facilities, and under staffing affected the accuracy of the monthly central morgue reporting of deaths, resulting in significant under estimates regarding fatalities. Minor short-term gains in accuracy can be achieved by conducting civic-action projects to upgrade and improve the morgue operations. Nonetheless, the morgue data was an essential source of information to track the effectiveness of ongoing operations that are aimed at curtailing sectarian violence.

Operation Bright Light

The history of Iraq is a history of the nomad. Arabs throughout the history of the land have been desert dwellers moving from one wadi to the next, looking for water and feed for their livestock. Eventually, nomads moving from the west came upon the Euphrates and the Tigris, settling along their banks and building the Iraq we know today. However, the sense of the nomad remains and the average Iraqi is very mobile. They travel on pilgrimages, for work, to visit friends and family, and when they are in danger.

Throughout 2006, MNF-I believed that movement of groups of Iraqis from threatened areas was an indicator of civil war. The assessment was that in fact it was not an indicator of civil war, but displacement was occurring to realign areas by religious affiliation. It was also an indicator that the Iraqi middle class was fleeing the country. The MND-B began an exhaustive effort to track and validate reports of displaced civilians. The highest numbers that the division could validate in Baghdad, Karbala, Babil and Najaf provinces was about 54,132. Out of a population base of 11 million, the division understood that 0.5 percent of the population was displaced primarily due to security reasons. As a comparison, New York City (which has a resident population of 8 million) has about 33,000 homeless or 0.4 percent of the population. Regardless, the MND-B took measures to prepare a contingency operation to conduct IDP (Internally Displaced Person) management.

During the course of July through September, Shi'a militias and political arms moved into Sab Al Bor, north of Baghdad, an area generally populated by Sunni families. Applying the Hezbollah model, the Office of Muqta a Sadr moved on the populace. The template was to influence the populace with charity, such as schools and clinics, and move them to the Shi'a side.[22] Visible signs of the Sadr movement incursion into Sab Al Bor were the posting of Muqta posters throughout the town.

In September, the Sunnis counterattacked. Shi'a pockets were mortared, forcing a displacement of some 3,000 Shi'a out of Sab Al Bor and into Kadhamiyah. On 7 October, MND-B through the 3rd Battalion, 1st Brigade, 6th IA Division confirmed the location of two IDP camps in Kadhamiyah. Camp 1 in Muhallah 425 (block number) had at least 1,200 people, and Camp 2 in Muhallah 427 had 600 people.

The Kadhamiyah District Advisory Council chair, Sheik Hassan, also verified that families had relocated into the area. His basis was the number of people registering for government allotment coupons, kerosene, propane, and food supplements.

The Sab Al Bor displaced persons presented death threat letters from the Omar brigade, targeting Shi'a for helping the Army of Muqta and the "infidels." The MND-B mission was to provide HA in order to mitigate the impact on the beladiyah (borough) of these displaced civilians and alleviate any human suffering. Specifically, the MND-B wanted to curtail any further escalation of displacement and as accommodations become more cramped, curtail any health issues.

Operation Bright Light went into the planning phase on 11 October and was implemented 3 days later. In order to keep the HA operation low key, coalition forces maintained minimal visibility and maximum support. Support services would be contracted out, and the Iraqi Army would be tasked for security and quality control on the services provided by the local contractors. Some of the important tasks were as follows:

a. Determine locations and quantities of relief supplies available.

b. Coordinate with the Iraqi Red Crescent Society.

c. Coordinate with the Iraqi Army to distribute HA.

d. Find local contractors to provide and deliver relief supplies such as generators, fuel, food, water, bedding, and latrine facilities. Additionally, provide milk, baby formula, diapers, and infant/family care items such as toys, soccer balls, and school supplies for an estimated 20 percent of the camp population.

e. Coordinate with Brigade Combat Teams to conduct reconnaissance, and atmospherics.
f. Conduct face-to-face engagements with local leaders.
g. Conduct MedOp with tactical human intelligence (HUMINT) teams and tactical PSYOP teams support.

The 2/1 AD attended the Kadhamiyah District Advisory Council meeting to confirm local government support and visited the two camps on 10 October. The commander of the 1/4 ID also visited the camps. A summary of the conditions follows:

a. Camp 1 had 736 people, and their one drinking water point had low pressure. One family lived in each tent, and sanitary facilities were minimal, consisting of one small building and one lavatory. The immediate needs of Camp 1 were sanitary facilities, food, drinking water, kerosene, and blankets.
b. Camp 2 had 600 to 700 people living and sleeping on the floors of the hallways and classrooms. Lavatory facilities were minimal. The immediate needs of Camp 2 are sanitary facilities, drinking water, food, cots, blankets, and kerosene.

The Iraqi Army and the Kadhamiyah District Advisory Council assembled and delivered relief packages on 14 and 15 October. Random checks of the area were scheduled during 16-21 October, and 2/1 AD and the 6th IA Division were to conduct a formal survey on 22 October. However, after the 1/4 ID commander returned from his visit to Kadhamiyah and met with the Sab Al Bor leadership, he closed down all civic-action projects in the area due to the attacks and the displacement.

By 18 October, 80 percent of the displaced persons had left the camps. By 22 October, Camp 1 was closed and by 28 October, Camp 2 was closed. The IDPs had returned to Sab Al Bor, the attacks had stopped, and it was safe to return. Later the civic-action projects restarted.

Operation Austin Light

The purpose of Operation Austin Light was to facilitate communications and cross-talk techniques and procedures for training and incident command between various Baghdad first responders. The City of Austin sent two deputy police chiefs and an assistant fire chief as trainers for the Rashid and Karhk Baghdad Fire Departments in incident management.

Figure 17. Operation Austin Light, 15–16 October 2006, Adnon Palace inside the Green Zone, Baghdad. Participants included the Karkh and Rusafa Fire Departments, the City of Austin Police Department, the City of Austin Fire Department, and MND-B.

The focus was establishing incident command and providing Iraqi emergency management services with the tools to become self-sustained and the foundation to work toward building a long-term capability for responding to emergencies:

a. Apply technical rescue capability.
b. Establish a perimeter control.
c. Suppress fire.
d. Provide incident command structure.
e. Provide secondary device sweep.
f. Establish a triage area.

The results were that representatives of the East and West Baghdad fire departments attended the exercise and left with a better understanding of the importance of coordinating the various emergency services.

Operation Holland Light

Student discontent with the government fosters an insurgency. In the Global War on Terrorism, radical students have in some cases become the Al Qaeda revolutionaries. Educated in madrassas across the world, religious students are filled with fundamentalist fervor to overthrow the apostate governments of the Middle East who have aligned with the U.S. Once indoctrinated, the students depart their madrassas to conduct jihad, and where they are found are the dead spaces. Rural areas provide safe havens for the insurgency. The enemy travels to and from the farmlands to the cities and conduct their attacks. They seek places in the countryside to hide themselves (using the hospitality of the farmer as they move from town to town) and weapons (in caches under palm groves and along irrigation canals). These are classic insurgent tactics.

In conjunction with the Brigade Combat Teams operating in rural areas, MND-B put together several programs to build civil cooperation in order to divulge the location and disposition of anti-Iraqi forces and the insurgency. If the enemy uses the agriculture belts for his staging bases, how do coalition forces separate the guerrilla from the populace?

After his arrival in Iraq, LTC Daniel Holland led the initiative for MND-B to establish a rural COIN program. On 18 May, his vehicle struck an IED, and he was killed instantaneously. Operation Holland Light was named in his memory and serves as an umbrella for the various agriculture development programs designed to win the farmers and defeat the enemy. The purpose of the operation was as follows:

a. Facilitate the gathering of information from farmers.
b. Execute short duration civic-action projects that have an impact on the populace and turn the people against the insurgency in order to reveal the enemy's location, disposition, and intent.
c. Build lasting ties between and foster civil cooperation with government officials.
d. Enhance the legitimacy and influence of the local and national governments.
e. Strengthen local and national economies.

To accomplish these tasks, the operation sought to facilitate the veterinary, agricultural, and medical efforts in concert with local officials. The effort was focused in the Abu Ghraib and Mahmudiyah Qadas areas with the goal of separating the local people from insurgency. Civil action was used to create a favorable environment in which to support current and anticipated operations and to gain trust and confidence of the people for their local and national government. Professionals known as *local nationals* were employed to facilitate the legitimacy and success of the operation.

After LTC Holland's death, LTCs Diana Brown and Neil Aihle took over and put these ideas into action. Several insights were developed from their experiences during this operation and others; major initiatives included veterinarian, agricultural, and MedOps:

a. Facilitate vaccination and treatment of animals by local veterinarians with a specific focus of maintaining the food supply for the populace.

b. Facilitate the clearing of drainage canals, distribution of fertilizer/seed, and standing-up of cooperatives. Cattle farms, chicken farms, and fish hatcheries appear to be the most economically viable agricultural businesses. The cattle are a cross breed of a "regular" Iraqi cow and a water buffalo. Chicken farms are mostly egg farms, and initial surveys identify that farmers requested veterinary assistance (vaccines and animal-husbandry services).

c. Coordinate MedOps with the Ministry of Health—the manning and supplying of medical clinics and hospitals to provide basic medical and minor dental care. MedOps are also combined operations where coalition forces treat local nationals for basic medical care only.

Major initiatives involved the agricultural associations (cooperatives) with objectives to improve the economy through the development of farm market forces as well as improve the quality and quantity of agricultural produce and products. The intent was to foster principles of cooperation and democracy in a hard-to-reach segment of the population—farmers.

The economy of Iraq was undergoing transformation to adjust to market-oriented reforms and move into the industrial era. The involvement of the Iraqi government and parastatals (state enterprises) in the Iraqi economy has been harmful in many ways. As a result, the government and parastatals need to be called upon to intervene less and to let markets work. In Iraq and especially the Abu Ghraib region, government policy has consisted of approaches that resulted in less, rather than more, effective agricultural operations and fostered isolation rather than cooperation in organizational structure and function. Farmers and agricultural participants in Abu Ghraib simply waited for governmental provision of agricultural products, then sold back all excess product to the government at a fixed price. Under this system, no cooperative organizations were ever created and genuine participation in a cooperative sense never occurred. As a result, members of the farming communities have, in many cases, become alienated from issues that should be within their control and from having influence on issues that should be of direct concern to them, such as the marketing and pricing of their own products.

The potential of genuine cooperative organizations to contribute to rural development based on popular participation has, to a large extent, been wasted. The very concept of cooperative self-help has been ignored or fallen into disrepute. Thus in Iraq, a review of policy concerning cooperatives was needed. The revised approach was based on the principles of individual participation and consultancy rather than governmental intervention. Existing policies need not be completely abandoned, however. In many cases, they can be adapted to suit a more participative and market-oriented approach.

During the time of Operation Holland Light, Iraqi farm income was projected at a $16.5 million annual loss. This calculation was primarily from lost productivity and lost local economic stimulus from associated activities as well as through direct farm-market commodity devaluation. Losses caused by a lack of development include inhibition to developing democratic principles, self-reliance, and self-sufficiency that cannot be measured in dollar value alone. The record however, clearly reflects that

Figure 18. Sheep-dip tanks

specific *policies* provided the seedbed to sow hatred and cultivate terrorist ideologies: isolation rather than cooperation and governmental dependency rather than independence combined with theocratic ideologies.

Encouraging agricultural cooperatives involved providing extension trainers to assist farmers with farm-market information. In addition, sheep-dip tanks placed in conjunction with agricultural co-ops contributed to parasite control and helped to improve meat and wool quality and quantity.

The plan also called for engaging the Ministry of Health and Ministry of Agriculture for administration and inspections of slaughterhouses as a steady source of safe and reliable meat for the local populace. A general food safety campaign was developed to install confidence in government institutions, employ people, and save lives. Additional facets of the program included a seed-purchase program and a revitalization of the poultry industry. In addition the operation sought to improve the capability of the Abu Ghraib Hospital, which had antiquated and marginal equipment and no maintenance-support program. At the time of this report, activities were underway to work with the Ministry of Health to improve services, equipment, medical stores, and staff.

By July, the religious violence had intensified to an alarming level. Sunni on Shi'a attacks had increased to the point that the focus of MND-B operations was interdicting the religious strife. Operation Together Forward II went into action. Using a "clear-build-hold" methodology, U.S. and Iraqi forces began a systematic operation to protect the populace "block by block." Chapter 6 takes a closer look at Operation Together Forward II, specifically the "build" phase designed to clean up and restore essential services and governance to large sections of Baghdad.

6. A Closer Look—Building Essential Services in Operation Together Forward II

In July, leaders decided that a new operation was needed in Baghdad to restore security and win the trust and confidence of its people. Coalition and Iraqi security forces began Operation Together Forward II in early August in coordination with the Government of Iraq, the Baghdad Provincial Council, the City of Baghdad (Amanat), and District and Neighborhood Advisory Councils. Substantial progress was made in providing essential services to meet the basic needs of 495,000 people in the focus areas. The effort added "build" within the clear, hold, and protect methodology for COIN operations.

The operation was carried out in three phases: clear, build, and hold. The 172 Stryker BCT established security in an objective area and conducted clearing operations for 4 days. Immediately after the clear phase, 2/1 AD, 1/4 ID, and 506 RCT began the build phase where trash pickup and essential service assessments were the initial big push. Once those operations were in order, the three units moved into the hold phase to work with the local government in establishing routines and continuous work, projects, and improvement in the neighborhoods.

The Plan

The civic-action plan conducted during Operation Together Forward II centered around three types of projects:

a. *High impact/short duration* was defined as civic action that takes from 1 to 90 days to complete. For Operation Together Forward II, the focus was on 7-day duration civic actions, specifically involving civic cleanup, civilian MedOps, building civil service, and improving public utilities. These projects are items that build the local government's ability to protect and provide for their citizens. They also reduce or remove leading causes of instability (i.e., electricity, medical, lack of governance, and trash). MND-B had direct management responsibility for these actions.

b. *Mid-term stability* was defined as civic action that would take between 90 to 180 days to complete and provide opportunities for future employment and include major war-damage rubble removal. These projects strengthen the local government's ability to provide for its citizens and expand their capabilities through development of ongoing and new civic actions. Here, MND-B took direct action with reinforcing support from other organizations.

c. *Long-term/economic development* involved civic action that lasts between 180 days to 3 years. For Operation Together Forward II, the focus was to revitalize the industrial base for market growth that would generate employment in and around the hot zones. Long-term/economic development ensures the local government can sustain and increase their capabilities and enable the local citizens to have a future that includes safety, economic security, and prosperity. The MND-B supported these activities with in-depth action provided by other organizations.

Sequencing the civic-action program was critical. The civic-action sequence of events for Operation Together Forward II aimed at stabilizing objective areas after strike operations. This involved cordon-and-search operations and civic-action projects that encompassed the fields of public administration, utilities, health, and safety. The critical tasks included the following:

a. Improve local government by staffing, training, and equipping the District Advisory Council(s).

b. Provide, improve, and upgrade the electrical services and consumer fuel distribution.

c. Conduct Civil MedOps (visiting doctors) for basic medical care to the citizens.

d. Initiate trash and debris removal, traffic control measures, and passive security.

The civic-action projects were designed to flow through the following cycle: assessments, security, and resources. Each Brigade Combat Team conducted initial surveys of their respective area of responsibility. The initial survey determined the immediate quality of life needs (i.e., food, water, utilities, and medical) that are required to bring the local population to a sense of "normalcy" as quickly as possible. The focus of the surveys was power substations and distribution points, water treatment facilities, sewage collection and treatment points, phone distribution centers, and cellular towers.

Security for the workforce became paramount. Throughout the sequence of events, security for the local nationals performing the contract work, as well as for coalition forces personnel, became increasingly important to preempt and prevent worker harassments and death threats. As Operation Together Forward II progressed, so did the intimidation. At one point, a work stoppage had occurred and an Iraqi Police battalion provided additional security.

Contracting for Labor

Numerous resource requirements were established during the planning phase that included internal and external needs. These resources were sourced prior to commencing operations. The goal was to use local contractors to the maximum extent possible with an emphasis on neighborhood or beladiyah-based contractors. Bringing laborers from outside a neighborhood sometimes had negative consequences such as intimidation and attacks against the workers. For example, the U.S. Army Corps of Engineers kept complaining that security problems (i.e., worker intimidation) were putting completion of the R3 Water Treatment Plant off schedule. The Corps was employing Sunnis from the Karkh side of Baghdad for the R3, which was located in a Shi'a area, adjoining Sadr City. The working intimidation was directly due to Sunnis imported to work in a Shi'a backyard.

For those Brigade Combat Teams that could not find contractors locally, MND-B found backup contractors to support the teams as needed. Some of the resource requirements follow:

a. Trash contractors to clean up/remove trash and debris to include damage created from previous operations
b. Electrical contractors for electrical wiring upgrade/repair/replacement
c. Water and sewer contractors for emergency repairs
d. General contractors for traffic control measures (i.e., stop signs, street lights, and security cameras)
e. General contractors to fix/refurbish a maximum of four fuel stations in each Muhallah and to provide dedicated fuels

(e.g., benzene, liquid petroleum gas, and kerosene) delivery/ distribution for each station to become fully functional

f. Technical assistance teams to provide quality assurance and quality control of all CERP projects in the area of responsibility.

In addition to civic-action projects, Operation Together Forward II fully engaged and involved the local governments for activities in the objective areas. This empowerment strategy helped build local government. By involving the Neighborhood Advisory Councils (NACs) and District Advisory Councils (DACs) in the project nomination and community interface, the populace saw that the projects were directly tied to these local leaders. Consequently, the provincial government (especially the city) began interacting with the DACs and NACs to synchronize local civic action with city-wide projects. Such was the case in Adhamiyah with Sheik Hassan, the DAC chair, who with LTC Hall and the Ministry of Electricity revamped 34 transformers and began to restring wire in an area that had not been supplied with power for 4 months.

Figure 19. Typical Electric Grid infrastructure in Baghdad

The Need for Security

Engaging the local government and the members of the communities has always been a civic-action task. During Operation Together Forward II, coalition forces ran into many problems of government officials not following through or showing up for meetings. In Ghazaliyah, Ameriyah, and Shulla, the main problem was always security. Workers did not want to come to work for fear of being killed. Government officials were kidnapped, murdered, and threatened when they were seen working with coalition forces, and even the beladiyah designated the areas as "red" zones and would not conduct repairs in the communities. Conversely, in Adhamiyah and Sha'ab/Ur, no problems existed with contractors showing up to work; they were actually able to employ on average 450 workers a day over a 3-week period, including Friday prayer day.

The focus of projects varied between the areas. 4/4 ID maintained a focus on sewage and water, while 506 RCT focused on HA and electrical problems. 2/1 AD worked on long-term projects and establishment of government services such as regular trash pickup, bank, and Iraqi Police station openings as well as long-term sewage

Figure 20. Local nationals employed for trash removal or civic cleanup projects

repairs. Each area's first task was to clean up trash, and each unit was successful with the initial effort. While difficult to change a mind-set, the units worked with local nationals to set a standard of cleanliness and to establish regular pickup as well as systems for removal.

Operation Together Forward II created short-term jobs that supplemented many of the families' incomes. With the school year underway, many of the schools were open, and the children were attending classes. The hope was that the security concerns would not increase because the coalition forces were in the area consistently. The markets were opening again, and good dialogue began between the coalition forces and the local governments. Not everything was good news in the objective areas, however. Many trash piles still existed, long-term jobs not yet established, and many of the essential services were still lacking due to contractors not working, the Ministry of Interior not providing maintenance, and many local nationals still living in fear. What follows are outlines of some of the projects in the objective areas: Doura and Risala; Adhamiyah and Sha'ab/Ur; and Ameriyah, Ghazaliyah, and Shulla/Nur.

The 4/4 ID Area of Operations: Doura and Risala

Operation Together Forward II started with clearing in Doura (August), Of the 9,082 homes searched, 92 doors, 192 locks, and 58 windows were replaced within 48 hours because broken or damaged during the clearing phase. The immediate action taken during the next phase was trash cleanup. Its impact on the people was evidenced by the increased activity in the Doura Market, which included the opening of new shops and vendors. Doura residents began to consolidate their trash to make it easier for the contractors to pick it up and keep the area clean, thereby demonstrating a sense of pride and commitment.

Figure 21. Muhallah 826 Mechanix Street cleanup, before and after

Coalition forces also worked with the Iraqi Army to distribute HA items to local nationals. Over 3,000 "meals in a box" and water were distributed throughout Doura and Risala. Waterline breaks were also repaired in Muhallahs 822, 824, and 826; and a large water-network renovation project occurred in Muhallahs 824, 826, 842, 804, 806, and 808.

In August, September, and October, MND-B employed 15,224 resident and nonresident workers for trash collection, sewer contracts, and general electrical work. The average number of daily workers in the Doura beladiyah totaled 146; in Risala, the number was 171. Workers removed 67,902 cubic meters of trash between 26 August and 31 October.

By November, the following projects were either ongoing or projected:

a. Repair Muhallahs 850 and 860 sewer network and the 824, which collapsed due to overuse and lack of power and maintenance

b. Improve the Al I'ilam, Al Salam, Tahaddi, and Hadar primary health-care clinics and the Aka School grounds

c. Improve and repair the Muhallahs 883, 885, and 843 electrical network; New Amel 11-kilovolt electrical feeder; and Muhallah 879 11-kilovolt electrical feeders
d. Repair Sewer-Pump Station TC3 and Muhallah 861 sewer and water
e. Add street lights on major roads and the Doura Market to provide better security
f. Repair (via Army Corps of Engineers) two gas stations to improve efficiency and effectiveness at fuel stations and deny a sanctuary for the enemy.

Figure 22. The Doura beladiyah (with vehicle) and marketplace (with metal frames)

Night letters were an indication that COIN operations were succeeding. The basis was that the enemy leadership had determined the populace was moving away from their influence and toward that of the Government of Iraq forces. The night-letter translation that follows demonstrated that a COIN was fought for control of the populace, not terrain, and the center of gravity is the people. MND-B patrol found the letter in Doura, and the G9 section translated.

> In the name of God the most merciful
>
> Thanks God and peace be upon Mohammed:
>
> Lately we have come to know that Mujahideen had received money from contractors in order to allow them conducting their works such as trash cleaning or constructing... etc. and we say thanks God we have not received any money from any contractor and whomever says so is a liar and soon will have his fate decided.

On the other hand, some hypocrites who call themselves as Mujahideen, they cheat people in this regard in giving warranty to the contractors, we shall let them regret, God willing. We do know them and we do not want to name them.

Last, the high committee of Mujahideen in Doura has decided refraining all works that run by contractors like trash cleaning which is under the supervision of DAC or Baghdad Municipality or contractors. We warn all who are working in this channel; supervisor or worker...etc. has to quit working otherwise he shall be our target; he either will be killed or kidnapped.

And we warn all Lorries, machine drivers from working with those people.

The invaders said through their media we will clean up Doura from trashes and terrorists and what they meant by terrorists is Mujahideen. Thus, Ya...Doura bravest do you accept this thing...we say to the Americans our sons will not work with you in this job and we order you the disbelievers, the invaders to clean up our neighborhood from trashes and we will clean it from the invaders, the sons of monkeys and pigs and sons of Iranians and God is making the believers the conquerors and he is good judger.

We repeat our warning once more to those who are working with contractors except (Potable water and Sewer) have to stop working right now otherwise, their fate is in doubt and will be sorry and God is our witness...and we will make them as a sample for the rest who do not take lessons.

God we have told and be our witness...and everybody should pass to the nonattendance God is Greatest...God is Greatest....God is Greatest

And the pride for God, his prophet and believers.

The 506 RCT/101 AASLT Area of Operations: Adhamiyah and Sha'ab/Ur

After clearing the buildings in each area, the initial effort was toward trash cleanup in the Adhamiyah and Sha'ab/Ur. Next, the power grids were addressed, ensuring that residents of each area would have power a minimum of 4 hours a day. Generator distribution to the NAC and DAC assisted the residents and contractors with supplying power during the down hours of prime power.

Figure 23. Muhallah 316, before (30 August) and after (6 September)

Table 5 summarizes activities accomplished in Adhamiyah and Sha'ab/Ur.

Table 5. Results of Civil-Action Activities

City	Activities
Adhamiyah	Cleared 11,659 buildings and replaced 1,203 doors, 1,260 locks, and 42 windows.
	Employed 18,827 resident and nonresident workers for trash collection, sewer contracts, and general electrical work in August, September, and October.
	Conducted three MedOps, treating 450 civilians (65 men, 157 women, 185 children, and 43 infants).
	Removed 47,511 cubic meters of trash between 26 August and 31 October.
	Delivered fifty 5-kilovolt generators.
	Distributed one hundred fifty 3-kilovolt generators to schools, the DAC, NAC, and businesses.
	Emplaced twenty 50-kilovolt generators in Muhallahs 308, 310, and 312.
	Repaired water/sewer breaks in Muhallahs 302, 312, 314, and 316.
	Transitioned responsibility for trash-collection services to the beladiyah on 10 October

Table 5. Results of Civil-Action Activities (continued)

City	Activities
Sha'ab/Ur	Removed 34,800 cubic meters of trash in September and October. Employed 1,900 local workers for the cleanup. Conducted three MedOps, treating 628 civilians (184 men, 221 women, 176 children, 47 infants) with typical ailments being abrasions, headaches, stomach aches, diarrhea, and infections. Distributed clothing and toys to local nationals. Cleared 36,835 buildings and replaced 480 doors, 862 locks, and 8 windows. Distributed one hundred thirty 5-kilovolt generators to the NAC. Distributed 3,970 HA meals and 2,000 school-supply packages. Delivered sixty five 5-kv generators to Ur NAC for distribution to schools and businesses on 8 October.

Examples of planned projects in Sha'ab/Ur and Adhamiyah included work on sewage pumps at lift stations, rehabilitation of the storm water network, and working with the Ministry of Health to provide supplies to public health clinics.

Figure 24. U.S. and Iraqi forces ongoing repair of major sewer and water point breaks in Sha'ab/Ur districts of Baghdad

The 2/1 AD Area of Operations: Ameriyah, Ghazaliyah, and Shulla/Nur

After clearing the buildings in each area, the initial effort was toward the trash cleanup, CM, and HA in Ameriyah, Ghazaliyah and Shulla. Long-term projects of fixing sewage and reducing IED threats were additional requirements within the neighborhoods. In the beginning of October, the NAC and DAC in Ghazaliyah began distributing 50 generators provided by coalition forces to local schools. Ameriyah was trying to reopen the Rafaidan bank that was closed due to a robbery in July. The Ministry of Finance acknowledged the need to reopen it but would not agree until security was no longer a problem.

Contractors did not want to work for fear of being shot or kidnapped. For example, four contractors were shot over the duration of Operation Together Forward II, and two front-end loaders were blown up and destroyed. The beladiyah and Amanat people were not willing to conduct their own projects and fix ailments within Ameriyah due to their area being designated a Red Zone because of security. Haji Aiyed, the most involved Ameriyah NAC chairman, was murdered in front of his family while shopping for HA in support of Ameriyah families during Operation Together Forward II.

The NACs in all three neighborhoods increased their willingness to get involved by working with the combined force on a regular basis. However, many times coalition forces had to push them to work with each other as well as deter militias and violence within their neighborhoods. A summary of results follows:

a. Cleared 39,307 buildings in Ameriyah, Ghazaliyah, and Shulla.
b. Employed 10,655 resident and nonresident workers for trash collection and CM between 15 August and 1 November.
c. Conducted three combined MedOps, treating 979 civilians (226 men, 368 women, 364 children, and 21 infants).
d. Removed 58,083 m^3 of trash between 26 August and 31 October.
e. Distributed fifty 3-kilovolt generators to schools (provided by coalition forces but distributed by the Ghazaliyah and Shulla NAC in order to promote the government system).
f. Transitioned responsibility for trash collection services in Ameriyah to the beladiyah in October.
g. Commenced USAID-funded vocational training 2 October in Shulla for 250 students with coalition forces working with USAID and the NAC for job placement for trainees.

Examples of other projects planned for Ameriyah, Ghazaliyah, and Shulla/Nur included the restoration of several gas stations in the objective area. Restoration was needed to decrease the amount of hours local nationals waited to receive their gas and help the owners regulate the traffic flow and address security concerns. Four new schools and refurbishment of 10 primary schools were scheduled. Efforts were underway to introduce the regular use of commercial dumpsters and residential trash bins. About 200 were purchased for Ameriyah, and USAID funded the purchase of dumpsters and trash bins for Ghazaliyah and Shulla. The bins and dumpsters were part of a working project with the beladiyah to help improve regular trash pickups as well as promote the use of bins rather than fields and streets within the neighborhoods.

Results of Operation Together Forward II

The $8.9 million civic-action program added the "build" within the clear, hold, and protect methodology for COIN operations. The results of the civic action follow:

a. Removed in excess of 221,300 cubic meters of trash from the streets of Baghdad.
b. Employed an average of 1,980 workers a day (59,326 total man-days) as of 1 November 2006.
c. Reduced unemployment by 4 percent Baghdad wide, which gained about 62,000 new jobs across the city.
d. Treated about 2,200 men, women, and children.
e. Began the electric grid and sewer rehabilitation for an estimated 495,000 Baghdad residents.
f. Built local government by empowering District and Nahia Council chairs and leaders with authority, gained from providing their citizens with essential services funded through the CERP.
g. Reestablished the University of Baghdad's technical assistance teams through a $2.1 million contract to employ upwards of 182 professors and students for conducting surveys, engineering assessments, and quality control/quality assurance of Baghdad CERP projects.
h. Established a comprehensive CM operation (Operation Strong Will) formulated in response to the 22 February Samarra Mosque bombing. The CM involved 15 HA and 15 construction

packages—prepacked bundles designed to provide immediate relief per pack for 100 individuals and light repair to 20 to 30 houses, direct access to the East and West Baghdad Fire Departments, and a comprehensive catalogue of 1,300 Sunni and Shi'a mosques in Baghdad.

i. Provided critical strategic synchronization for interagency operations that enabled the fusing of the economics, governance, and security lines of operations in the center of gravity through the allocation of two Civil Affairs companies to the Baghdad and Babil provincial reconstruction teams.

j. Established the Joint Planning Commission, an interagency, multiorganization synchronization board to coordinate $2.2 billion in reconstruction projects in the MND-B area of responsibility.

A key lesson from the operation was the imperative that civic action must follow strike operations. Such civic action must encompass public administration, public health, public safety, and public utilities. While units can identify local trash, electrical, water, and sewer emergency repairs, having general contractors do the work gets locals involved and provides jobs.

Continued initiatives are needed to keep up the momentum of Operation Together Forward and similar operations. The Government Service Office program is necessary to provide basic training for those administrative personnel that support government services. Iraq's government offices do not have trained staffs to operate offices efficiently. A prime example is traffic tickets. Are laws clearly written? What are the penalties for violators? Is there a fine? How do police write tickets? If the police write a ticket, where does the recipient pay it? Who processes the information to keep track of it?

Working through the University of Baghdad, a National Teachers' Corps is needed to establish a program to accredit all levels of teachers. Included in this package should be graduating teachers' salaries and job placement. Another required initiative is the National Library, with the recommended location being the University of Baghdad. Its design would restore Baghdad to its history of being a center for culture and knowledge. A new facility could serve as the national archives for the Government of Iraq, and it would give Iraqis a great sense of pride.

Another project is the revitalization of the Army Canal on the east side of Baghdad. Because Iraqis take great pride in parks, this project will turn a long stretch of ugliness into a beautiful walkway and waterway lined with trees. It will be an area where families can enjoy the day or simply go for a walk. These projects and other long-term initiatives have the potential to turn Baghdad into a livable city that will one day be a major center for the Middle East.

7. Conclusions and Insights

> All they saw before was American patrols ... up and down the main streets. Now we've got American patrols out there with the Iraqi Army. And we've got garbage trucks out here and workers from Amariya picking up garbage. It gave them a sense of security that things are a little better.
>
> — LTC Gian Gentile

The Battle of Baghdad was fought and continues using a combination of strike operations and civic action. The enemy morphed constantly over the course of 2006, but what we found was the emergence of a hidden enemy that surfaced when the time was appropriate. A variety of undercurrents were present:

a. Ebb and flow from the formation of an Islamic state to the establishment of a democratic republic of Iraq
b. Pro-western versus traditionalism
c. Sunni versus Shi'a to regain control of Baghdad.

Persian against Arab, or Al Qaeda and Sunni extremist against Shi'a Ayatollah and militia, regardless of the enemy situation or force, the civic-action plan sought to stabilize the 75 to 80 percent of the population who were the focus of the fight.

The MND-B civic-action program started reversing 30 years of neglect when Saddam Hussein tyrannized Iraq. During that time, the wealth of Iraq was spent on weapons and war instead of caring for the citizens or maintaining the infrastructure and improving the quality of life for the ordinary citizen. Without the capital improvements to the society, MND-B faced significant infrastructure challenges to repair and restore in the MND-B area of responsibility, everything ranging from road repair to revitalization of the decrepit electric power grid to refurbishing schools.

The civic-action program assisted in improving the quality of life for the villages, towns, and neighborhoods of Baghdad city and the surrounding districts, which involved $138 million total obligation for fiscal year 2006, with $75 million as the obligation target for fiscal year 2007. At the transition of authority, the 1,044 projects underway spanned nine categories; see Table 6.

Table 6. Civil-Action Program Overview, Accomplishments

Category	Percent	Synopsis
Water/sewer	14	146 projects, including 13 sewage pump stations, 19 water treatment plants, 6 compact water units, 26 sewer and water main breaks, replacement of 25 kilometers of water pipe, overhaul of canal cleaning, potable water delivery, and water distribution for 38 city blocks
Electric	10	107 electric network, grid rehabilitations, and power generations projects
Education	13	140 projects to include the renovation of 5.5 percent of Baghdad's 93 (of 170) schools and upgrading facilities for 111 elementary schools (310,800 students)
HA/MedOps	7	71 operations or projects
Health care	5	57 MedOps, including 21 hospital and clinic renovations
Civic cleanup	10	102 projects that removed about 260,000 cubic meters of trash and rubble
Transportation	14	145 primary and secondary roads repaired or paved, as well as sidewalk and curb repair/construction and bridge construction
Telecommunications, law/governance, economics/finance, agriculture/irrigation, food distribution	9	37 agriculture projects that helped farmers across Baghdad, Karbala, Babil, and Najaf provinces 14 Qada, District, and Nahia Councils building renovations 38 local government support and sustainment programs During peak employment, provided jobs and job opportunities for 53,908 local nationals (22,491 with CERP, 29,168 reduced unemployment, and 2,249 for Operation Together Forward II)
Other	18	187 projects—e.g., veterinarian operations, HA, school supplies, gas station repair, and CM

The solution in Iraq involves developing a fully resourced Internal Defense and Development (IDAD) plan of action. It would be a full interagency effort to bring all the elements of U.S. national power into play. The effort would be assisting the Government of Iraq and the Iraqi people in modernizing their country through the revitalization of their public administration or civil service, public health system, public safety, and public utilities. As advised by Joint Publication 3-07.1, *Joint Tactics, Techniques, and Procedures for Foreign Internal Defense (FID)*, 30 April 2004:

> The IDAD strategy is the full range of measures taken by a nation to promote its growth and to protect itself from subversion, lawlessness, and insurgency. The strategy focuses on building viable political, economic, military, and social institutions that respond to the needs of society. Its fundamental goal is to prevent an insurgency or other forms of lawlessness or subversion by forestalling and defeating the threat and by working to correct conditions that prompt violence. The government mobilizes the population to participate in IDAD efforts. Thus, IDAD is ideally a preemptive strategy; however, if an insurgency, illicit drug, terrorist, or other threat develops, IDAD becomes an active strategy to combat that threat.

With this writing the battle for Baghdad continues apace with many challenges for the ground troops. As military and civilian planners look ahead, they will need to address the causes of instability listed in Table 7.

Table 7. Baghdad Specific Civic Action

Cause of Instability	Civic Action to Remove or Reduce
Concern for being a victim of indiscriminate firing by American soldiers	Enforce disciplined adherence to rules of engagement, good fire-control measures, sighted in weapons
Feeling unsafe from attacks	Conduct security operations with vetted Iraqi forces, Army in active overwatch, police on the beat
Desire for Islamic nation, one that is built on a community of the faithful within Islam	Public-information campaign on tolerance and treating each other as one would want to be treated themselves
Unstable electricity—wants free service optimally, but will pay for 24-hour electricity; requires an even distribution of power	Revitalize beladiyah electric networks, along the Adhamiyah model
Ongoing car bombings and other insurgent attacks	Mobile checkpoints, none fixed along the 7 main and 20 secondary roads into and out of Baghdad city

Table 7. Baghdad Specific Civic Action (continued)

Cause of Instability	Civic Action to Remove or Reduce
Wanting respect from American forces	Add to the rules of engagement the standards of conduct to local nationals, with equal opportunity advisors as teachers/battle trackers
Closed universities, colleges, and schools; wanting them open with a tolerant syllabus, not fundamentalist	Focus on renovations, build a Baghdad University Teachers Corps, and teach tolerance and science
Iraq oil industry not self-sufficient throughout, thereby unable to generate revenues for the citizens	Focus reconstruction efforts on revamping the oil infrastructure first
Desire for honorable employment—more of a technical nature and white-collar perspective; wanting dirty work done by others	Create an Iraqi Job Corps to teach technical jobs, reinstate the civil service, train teachers, and hire third-country nationals as laborers
View of American forces as occupiers, not as liberators	Public-information campaign on how the coalition is assisting the citizens of Baghdad
Lack of local medical care if they are caught up in any violence	Pressure the U.S. Army Corps of Engineers to finish the primary health-care clinics and move completion to a priority behind grid revitalization
American arrogance (vs. assistance)	Embed Iraqi media with U.S. and Iraqi forces and personalize the American soldier to Baghdad citizens
Ongoing kidnapping and robberies	Place Iraqi Police on the beat, 24 hours a day, and a 911 system per beladiyah under a DAC security coordinator
Trash accumulation, dirty streets, and streetlights not working	Remove trash, transition contracts for U.S. control to DAC control, institute a service fee per household
Lack of communication from the Americans—want to know what is going on with security and infrastructure development	Hold a weekly TV and radio talk show, with a focus per belayditah to provide the citizens access to coalition forces
Lack of clean water in the homes	No new starts—complete all ongoing CERP projects pertaining to infrastructure and place a 3-year service contract on all

Table 7. Baghdad Specific Civic Action (continued)

Cause of Instability	Civic Action to Remove or Reduce
Desire for relaxing environments for families	Give the lead to the City of Baghdad for soccer field, parks, and outdoor amusement projects
Desire for a national identity	Baghdad-identity public-information campaign that recounts the glory of the past, the promise of the future

The final insight for civic action centers on the three driving principles: simplicity, results, and focus. MND-B was effective in keeping the civic-action program simple. Upon receiving direction from higher headquarters to embark on complex Baghdad-wide infrastructure development programs, which by U.S. standards was complex and by Iraqi reality impossible, we stayed with basic local needs.

By keeping efforts simple, results are achieved. In Iraq the populace is in Maslov's basement, and the focus is on the basic needs of the populace. To win the trust and confidence of the populace, you simply have to produce. Promises made must be kept. One who walks into a neighborhood and promises to fix all the problems and then not deliver is viewed as just another good-hearted American who cannot produce.

The principle of focus is pinnacle to the success of civic action—that is, analyze what the people need, not what we want to give them. Focus on *their* standards, then on many short-term civic-action projects spread over a wide area, not one major project spread over a long term. Also focus on areas in need. By doing so, the battle can be won by applying civic action, block-by-block.

Endnotes

1. *Mujahideen* (the Arabic literally meaning "strugglers") is a term for *Muslims* fighting in a war or involved in any other struggle. Mujahid and its plural, mujahideen, come from the same Arabic root as *jihad* ("struggle"). The word is the plural form as literally translated from Arabic meaning "struggler." In Islamic scripture, the status of *mujahid* is unequal to *qaid*—one who does not join the jihad. Available from http://en.wikipedia.org/wiki/Mujahideen (accessed July 2007).
2. The Joint Staff, 23 February 2007 briefing on Countering Ideological Support for Terrorism.
3. Tim Green, *The Red Zone* (New York: Warner Books, 1998).
4. Ian Frazier, "Invaders, Destroying Baghdad," *The New Yorker*, 25 April 2005.
5. Roger J. Spiller, *Sharp Corners: Urban Operations at Century's End* (Fort Leavenworth, KS: USACGSC Press, 2001).
6. Adrian Bogart, *One Valley at a Time*, JSOU Report 06-6 (Hurlburt Field, FL, August 2006), pp. 19-22 and 24-26.
7. Under Article 22 of the League of Nations Covenant, Iraq was formally made a Class A mandate entrusted to Britain. The award was completed on 25 April 1920 at the San Remo Conference in Italy. As news of the mandate reached Iraq in late May, a group of Iraqi delegates met with the British administrator and demanded independence. Wilson dismissed them as a "handful of ungrateful politicians." During the Holy Month (Ramadan), joint Sunni-Shi'a meetings were held appealing for unity against the British. On 30 June 1920, insurrection against the British erupted in Iraq. Over the next 6 months, the British lost 1,654 men and spent six times as much as they had spent on the entire wartime campaign in the Middle East.
8. Quote is from the poem "To a Louse" by Robert Burns.
9. See www.terrorism-news.blogspot.com/2005_10_02_archive.html (accessed 31 August 2007).
10. Ellen Knickmeyer, "Baghdad Neighborhoods Hopes Dimmed by the Trials of War," *Washington Post,* 27 September 2005, p. 1.
11. Kuffar, pl. Kaffir—person who refuses to submit himself to Allah (God), a disbeliever in God.
12. See http://video.google.com/videoplay?docid=908368152252752624 2 (accessed 31 August 2007).
13. Iraqi Advisor Task Force (IQATF) local national advisors (LNAs), October 2006

14. See www.tkb.org/Group.jsp?groupID=4575, MIPT Terrorism Knowledge Base (accessed 31 August 2007).
15. In western terms, *reconcile* is to resolve differences, but in Iraq, it is to define tribal and religious areas of control.
16. CPAC report—Surprise Passage of the Federal Region Formation Law by John Houston and Hussam Al Odeh, cultural and political advisors, 11 October 2006.
17. Joint Publication 3-07.6, Joint Tactics, Techniques, and Procedures for Foreign Humanitarian Assistance, 15 August 2001.
18. The Iraqi society is nomadic. The movement of civilians across the country is normal and not an indicator of civil war, as would be construed in a western construct.
19. The term *reconstruction* is a gross misnomer because used to cite repair, rebuild, or replace damage created by war. The U.S. prided itself on limited war damage. A testimonial is seeing the Special Republican Guards complex leveled in April 2003, but the mosque next to it without even dust. By the end of 2006, however, the destroyed building was still not rebuilt. Reconstruction really meant *infrastructure development.*
20. For the 16 governmental functions, see FM 41-10, *Civil Affairs Operations*, February 2000.
21. Ellen Knickmeyer, "U.S. Plan to Build Iraq Clinics Falters, Contractor Will Try to Finish" Washington Post Foreign Service 3 April 2006, p. A01; available from www.washingtonpost.com/wp-dyn/content/article/2006/04/02/AR2006040201209.html (accessed 13 August 2007).
22. The schools, clinics, and projects that Sadr would cite as his contribution to the local community were in fact civic-action projects from Multi-National Division, Baghad. Likewise was the case with Sunni leaders such as Saed Janabi who claimed that the cleanup and neighborhood revitalization of Doura was his 'political' effort (interview during Ramadan at the Deputy Commanding General, Support visit).

References

10th Armored Division Command. 27 March 1995. Translation of *SIAQAT AL=ML ALACBT? 131-3-12. 239* (FSTC-HT-0393-93, Division Standard Operating Procedures, Guerrilla Warfare Manual). Page FM8531. Language, Arabic. Country, Iraq. Country of origin, Iraq.

Akehurst, John. 1982. *We Won a War: The Campaign in Oman 1965-1975.* United Kingdom: Michael Russell Publishing Ltd.

Blanchard, Christopher. 20 June 2005. *Al Qaeda: Statements and Evolving Ideology.* Congressional Research Service (CRS) Report for Congress. Middle Eastern Affairs, Foreign Affairs, Defense, and Trade Division. Washington, DC: The Library of Congress. Available from www.law.umaryland.edu/marshall/crsreports/crsdocuments/RL3275906202005.pdf (accessed 23 August 2007).

Center for International Issues Research. 6 December 2005. Al-Qaeda Offers Revised Version of its Phased Transnational Strategy, Global Issues Report.

Combined Arms Research Library (Fort Leavenworth, KS). 1972. FM 31-23, *Stability and Support Operations.* Washington, DC: Department of the Army.

Cordesman, Anthony. 18 November 2005 (updated draft). "Iraq and Foreign Volunteers." Arleigh A. Burke, Chair in Strategy, Center for Strategic and International Studies. "Iraqi and Foreign Volunteers." Available from www.csis.org/component/option,com_csis_press/task,view/id,1478/ and www.csis.org/media/csis/pubs/051117_iraqforeignvol.pdf (accessed 31 August 2007)

DEBKA-Net-Weekly. 25 February 2004. "Al Qaeda Builds a Euro Army." Available from http://www.debka.com/article_print.php?aid=794 (accessed 23 August 2007).

Department of the Army. 29 August 1986. FM 90-8/MCRP 3-33A, *Counterguerrilla Operations.* Available from www.smallwarsjournal.com/documents/fm90-8.pdf (accessed 23 August 2007).

Department of the Army. December 2006. FM 3-24/MCWP 3-33.5, *Counterinsurgency.* Available from www.fas.org/irp/doddir/army/fm3-24.pdf (accessed 23 August 2007).

Department of the Army. October 1967. FM 31-73, *Advisor Handbook for Stability Operations.*

Faqih, Saad Al. 2 June 2005. "Al Qaeda: In Decline or Preparing for the Next Attack? An Interview with Saad Al Faqih." Interview by Mahan Abedin (London). Available from http://jamestown.org/terrorism/news/article.php?articleid=2369721 (accessed 20 August 2007).

Hossack, Andrew. 5-8 April 2004. *Historical Analysis of Terrorist Campaigns, With observations on current operations in Iraq.* Dstl/CP10135 Power-

Point presentation to Cornwallis IX. Ministry of Defense, Dstl Policy and Capability Studies Department.

Katzman, Kenneth. 17 August 2005. *Al Qaeda: Profile and Threat Assessment.* CRS Report for Congress, Middle Eastern Affairs, Foreign Affairs, Defense, and Trade Division. Washington, DC: The Library of Congress. Available from http://www.fas.org/sgp/crs/terror/RL33038.pdf (accessed 23 August 2007).

Macgregor, Douglas. 21 April 2005. *Transformation and the Illusion of Change: Where is the Army really headed? What is to be done?* Presentation to House Armed Services Committee. Available from www.thedonovan.com/archives/historystuff/IllusionOfChange(Macgregor)%20(2).pdf (accessed 23 August 2007).

McCaffrey, Barry. 18 July 2005. Memorandum for Senate Foreign Relations Committee. Adjunct Professor of International Affairs, U.S. Military Academy. Trip Report to Kuwait and Iraq, 4-11 June 2005. Available from www.foreign.senate.gov/testimony/2005/McCaffreyTestimony050718.pdf (accessed 23 August 2007).

Middle East Media Research Institute (MEMRI). 19 October 2005. *The Global Islamic Media Front: Al Qaeda in Iraq Has Replaced the Mother Organization in Afghanistan and is Spreading Jihad Throughout the World.* Available from www.terrorism-info.org.il/malam_multimedia/html/final/eng/eng_n/memri_10_ 05_e.htm (accessed 23 August 2007).

Molnar, Andrew. 1966. *Human Factors Considerations of Undergrounds in Insurgencies.* Department of the Army Pamphlet 550-104, Washington, DC: American University.

Schweitzer, Yoram, and Sari Goldstein Ferber. 8 November 2005. *Al Qaeda and the Internationalization of Suicide Terrorism.* Tel Aviv University: The Jaffee Center for Strategic Studies. Available from www.tau.ac.il/jcss/memoranda/memo78.pdf (accessed 29 August 2007).

Spiller, Roger. 2000. *Sharp Corners: Urban Operations at Century's End,* Fort Leavenworth, KS: U.S. Army Command and General Staff College Press.

United Nations Security Council. 8 June 2004. *Resolution 1546.* Available from www2.unog.ch/uncc/resolutio/res1546.pdf (accessed 23 August 2007).

United States District Court Southern District of New York, United States of America Against Usama bin Laden, et al., S7R 98 Cr. 1023. 13 February 2001. New York. Available from www.globalsecurity.org/security/library/policy/national/05-06625.pdf (accessed 23 August 2007).

Woodward, Joel. 2002. "A Comparative Evaluation of British and American Strategy in the Southern Campaign of 1780-1781." Master's thesis, Army Command and General Staff College, Fort Leavenworth, KS. Available from www.ntis.gov/search/product.asp?ABBR=ADA407971&starDB=GRAHIST (accessed 23 August 2007).

Yaphe, Judith. 2003. "War and Occupation in Iraq: What Went Right? What Could Go Wrong?" *Middle East Journal*, Vol. 57, No. 3 (Summer). Available from www.mideasti.org/pdfs/Yaphe381-399.pdf (accessed 23 August 2007).

Yitzhak, Nakash. 7 February 1994. *The Shi'is of Iraq*, Princeton University Press.